WHO LOSES WINS

Winning Weight Loss Battles

A 'Fat Mentality'
Versus
A 'Fit Mentality'

Dr. David H. Dighton

Who Loses Wins. Winning Weight Loss Battles. The 'Fat Mentality' versus The 'Fit Mentality'.

A CIP catalogue record for this title is available from the British Library

ISBN: 978-1-7385207-1-8

Copyright © 2024 by Dr David H. Dighton. First edition 2024

All rights reserved. No portion of this book may be reproduced in any form without written permission from the publisher, author, or under licence from the Copyright Licensing Agency Ltd.

Published in the UK by MediCause, 115 High Rd., Loughton, Essex IG10 4JA UK

www.daviddighton.com   www.medicause.co.uk

This publication is designed to provide accurate and authoritative information in regard to the subject matter covered. It is sold with the understanding that neither the author nor the publisher is engaged in rendering legal, investment, accounting or other professional services. While the publisher and author have used their best efforts in preparing this book, they make no representations or warranties with respect to the accuracy or completeness of the contents of this book and specifically disclaim any implied warranties of merchantability or fitness for a particular purpose. No warranty may be created or extended by sales representatives or written sales materials. The advice and strategies contained herein may not be suitable for your situation. You should consult with a professional when appropriate. Neither the publisher nor the author shall be liable for any loss of profit or any other commercial damages, including but not limited to special, incidental, consequential, personal, or other damages.

Cover by Stefan, P. @pwaperpro & fiverr.com

Other illustrations by Dr David H. Dighton

MEDICAUSE

MediCause

www.medicause.co.uk

About the Author

Dr. David H. Dighton qualified at the London Hospital Medical College in 1966 with MB and BS (London) degrees. In 1970, he became a British Heart Foundation Fellow in Cardiology at St. George's Hospital Hyde Park Corner, London, working with cardiologists Dr Aubrey Leatham and Dr Alan Harris. From there, he published scientific papers relating to cardiac pacemaker function. In 1973 he became a MRCP(UK), and later became a Lecturer (London University) in General Medicine and Cardiology at Charing Cross Hospital, London.

In 1980, he was appointed Chef de Clinique (Assistant Professor) at the Vrije University Hospital in Amsterdam. Having returned to the UK in1982, he worked both in his own private cardiac practice in Loughton, Essex, and at the Wellington Hospital, London. In 2000 he started a private diagnostic cardiac centre specialising in heart disease prevention and the early detection of heart and artery disease.

He retired from practice after battles with the medical regulators he disagreed with (see his book: The NHS. Our Sick Sacred Cow).

Other Books by the Author

Eat to Your Heart's Content. The Diet and Lifestyle for a Healthy Heart.(2003). HeartShield Ltd. ISBN: 0-9551072-0-2 (Hardback).

HeartSense. How to Look after Your Heart.(2006). HeartShield Ltd. ISBN 0-9551072-1-0. (Paperback).

The NHS. Our Sick Sacred Cow. Causes and Cures (2023) (Paperback & ebook). ISBN 978-1-3999-6027-4

How to Become Heart-Smart. A User's Guide to Heart Health & Heart Disease Prevention. (2024). (Paperback & ebook). ISBN: 978-1-7385207-0-1

In Preparation:

The Doctor's Apprentice. The Art and Science of Medical Practice. Clinical Cardiology for Students.

CONTACT

For more information, go to 'www.daviddighton.com' or email: david@daviddighton.com or daviddighton@loughtonclinic.org

Acknowledgements

Had it not been for my patients, I would never have been aware of the problems faced by those who repeatedly try to lose weight.

I am indebted to Jean Raven, Kim Yardley, Kathy Hall, Barbara Bird, Janet Westbury and may others, kind enough to contribute their points of view.

My thanks go in anticipation to anyone prepared to offer constructive comment.

DEDICATION

This book was inspired by the memory of a special friend:
Ralph Barton
(1942 - 2001)

With wit and wisdom;
Love and laughter,
Fun and insight;
He brightened every gloomy day,
Leaving us,
Wanting more.

He bypassed life's quick-sands;
Rode its roller-coasters;
Juggled its stresses.
And from their essence,
Concocted comic lessons,
As gifts.

Then, softly one night,
Came two conspirators:
Obesity and heart disease;
Taking him by stealth;
Leaving us bereft,
And forever diminished.

No weak cards,
Were his to play:
Neither vanity nor pride;
Selfishness or conceit.
For those he loved,
Their joy was his only desire.

Would that I could have saved him
For more than a day:
To replace our mourning tears
With those of joy,
Again to take his gifts of pleasure;
Again to count him as a treasure.

List of Tables

Table 1: Weight of Food in Grams and Ounces Containing 100 Kcals.

Table 2: Portions of Food containing 100 Calories

Table 3: Food containing most sodium salt.

List of Figures

Figure 1: The Dieting Pyramid.
Figure 2: The Dieting Pyramid. Appetite Control
Figure 3: The Food / Appetite Cycle.
Figure 4: Appetite Progress with Time.
Figure 5: The Dieting Pyramid. Portion Size Control
Figure 6: The Dieting Pyramid. Carbohydrate Reduction.
Figure 7: Find Your BMI.

CONTENTS

Introduction	1
PART ONE WEIGHT WORLDS	8
Chapter 1 Two Cultures: A 'Fat Mentality'. A 'Fit Mentality'	9
Chapter 2 Is Yours a 'Fit' or 'Fat Mentality'?	21
Chapter 3 What Mentality Controls You? Take the Test	29
PART TWO DO YOU HAVE A WEIGHT PROBLEM?	42
Chapter 4 What Weight Problem?	43
Chapter 5 Who Said You Are Overweight?	54

Chapter 6   79
Losing Weight. How Simple Is It?

Chapter 7   91
I'm Overweight. Whose Fault Is It?

PART THREE   102
THE DIETING PYRAMID

Chapter 8   103
Climbing The Dieting Pyramid

Chapter 9   116
The Second Step. Appetite Control.

Chapter 10   133
Food Quantity Control.

Chapter 11   154
Food Choice: Cutting the Carbs

Chapter 12   167
Food for a Healthy Heart and Circulation.

Chapter 13   175
Mantras to Defeat a 'Fat Mentality'

Chapter 14   180
A 'Fit Mentality' is not for Everyone.

Chapter 15   187
Eating Habits, Eating Styles & Culture

PART FOUR   201
ARE WE WHAT WE EAT?

Chapter 16 How to Exercise More	202
Chapter 17 Are You Disciplined Enough for Eating Less & Exercising More?	209
Chapter 18 Weight Loss: Truth & Myth. The Appliance of Science.	216
Chapter 19 Self Help and Medical Help	230
Chapter 20 Rounding – Up. Slimming Down.	245
Epilogue	248
PART FIVE APPENDICES	250
Appendix 1 What Should You Weigh?	251
Appendix 2 Overweight and out of breath? Is your life at risk?	254
Appendix 3 100 Calorie Food Portions & Carbohydrate Content	256
Appendix 4 Foods that Might Promote Artery 'Furring'	263
Appendix 5 Foods Likely to be Good for Arteries and the Heart	266

Appendix 6 272
Salt (Sodium) Units in Foods

Bibliography 280

Index 285

INTRODUCTION

After the experience I had as a cardiologist, occasionally directing a weight loss clinic, I decided to write this book to help those who have repeatedly tried to lose weight and failed. They may have found some diet and exercise regimes unsuitable. Weight loss regimes, like all medical interventions, work best when individualised. For those who prefer to choose their own methods, I have detailed the key facts, the latest science, and some tips and tricks to choose from.

The most useful issues concern appetite and food portion control, the role of carbohydrates in food, the contribution of our genes and gut hormones, and medical interventions. As important, is the recognition and management of the mindset that drives our eating, exercise and self-gratification.

Those who cannot control their self-gratification, and eat regardless of need, often have a self-indulgent mentality (a 'Fat Mentality'). Until they give precedence to a fitter mentality (their 'Fit Mentality'), they will not easily achieve sustained weight loss.

Many overweight people are happy with their weight. They have that right, even if medical professionals think unwise.

Some find it easy to lose weight by eating less and exercising more. They are likely to be those who are fit and disciplined by nature. This simple advice will work well for many, but for those who try repeatedly to lose weight, more detail can help. This applies especially to those with a disabling heart, lung, or joint condition. They will need the medical help detailed within.

The calories we eat, and the calories we use when exercising, are the focus of every book on dieting and weight loss, but typically, one key topic is missing. Like an undisclosed elephant in a room, an important topic is often ignored. The topic is a psychological one; namely, what drives some to become overweight, and others to be slim, fit and athletic.

Psychologically, we all have a 'Fat Mentality' and 'Fit Mentality'. These two components of our eating and exercise mindset co-exist, but are forever in opposition. The 'Fat Mentality' promotes our inactivity, self-indulgence and weight gain; our 'Fit Mentality' drives our desire for activity and the disciplined control of everything we do. Because they direct what we eat, and what exercise we choose to do, they strongly influence our body weight. A better understanding of this influence is crucially important to those who repeatedly fail to lose weight.

There are other important factors at work. In some cases, our inherited genetic profile can make successful slimming difficult. Others need to know what exercise is best, and how to control their appetite, their portion sizes, and the amount of carbohydrate they eat. I have introduced something new to this discussion: which foods are likely to promote the health of our arteries and benefit the heart.

Our character and personal values develop as we experience life. As we grow, we decide what is important to us, and what value to place on money, food, and relationships. These values form part of our outlook and mentality and direct our choices and behaviour. Subconsciously, they direct our choice of lifestyle. Some of us opt for decadence and an easy life (features of a 'Fat Mentality'); others opt for a life of endeavour and achievement (a 'Fit Mentality'). Our mentality, or mindset, never rests. It is active as we shop, when choosing the food we want to eat, and when deciding whether or not to exercise.

Our mentality plays an important role in all of our health and disease management issues. In matters of weight control, our attitude to eating and exercise needs to be considered alongside food calories, physical activity, thyroid and other hormones, our genes, our gut hormones and gut biome. To achieve weight loss success, we cannot ignore our mentality, simply because it can control all of our choices.

I first started writing this book when I was physically inactive, and not at all concerned about my weight. In February 2005, I was admitted to hospital with acute pancreatitis, one year after starting to write this book. Since then, the understanding of obesity and weight control has advanced. The most important and useful advances are discussed within.

My recovery took some time, even for the fit person I was before my illness. Prior to this, I had trained in a gymnasium for decades (four times every week without fail), doing what for me, was maximal weight and circuit training.

Having lost weight with a severe illness, I sat around, eating whatever I wanted, and doing very little exercise. I had written two books before getting ill, sitting at a computer, and doing no more exercise than it took to walk from my car to my bed. My take-it-easy lifestyle did little to encourage my return to physical fitness.

I quickly changed shape, putting on 14 pounds in a few months. That's not much, you might think, but underlying it, I recognised a fundamental problem. My mindset had changed. Apart from becoming disinterested in exercise, I couldn't stop eating, despite knowing what extra calories and carbohydrates (carbs) would do to me. I had a problem with my mentality. It had changed from a 'Fit Mentality' to a 'Fat Mentality'. While running weight control clinic sessions, I came to realise that many patients shared my problem.

Some overweight people who regularly fail to lose weight, will benefit from medical help. Appetite suppression, metabolism manipulation, psychological intervention and surgery, all have their place for those who suffer from a medical condition like heart disease, diabetes or infertility. Medical interventions are most likely to become necessary for those with a genetic profile that makes dieting and exercise ineffective.

Overweight people have many reasons for slimming. Some reasons are laced with hope; others are lightly sprinkled with fairy dust. Among those I have heard are:

- A desire to be fashionable.

- A desire to get back into those old clothes (back to a younger self).

- A desire to feel more comfortable.

- A desire to feel less breathless.

- A desire to look good in mirrors.

- A desire to improve health and reduce health risk.

- A desire to improve self-esteem.

- A desire to please others.

- A desire to remove bullying.

- A desire to remove the arm twisting of well-wishers, including that of a well-intentioned medical profession.

Eating occupies an important place in the rituals of every culture. The culture into which we are born, not only envelopes and nurtures us, it influences our eating and other habits for life. Our individual culture, influenced as it is by our family and social traditions, shapes our mindset.

Newly discovered forces are at work that cause some to put on weight more easily than others. Among them are genes, gut hormones and our gut biome, neurochemicals in our brain, our metabolism, the stresses upon us and our psychology. All play their part in what we choose to eat, how much we eat, the exercise we do, and the weight our scales will register.

Everyone fighting a lifelong battle to lose their excess weight will need to find which of the many factors I have detailed within, are most acceptable and relevant to them. Having first taken control of their 'Fat Mentality', their best chance of losing weight will come from controlling the calories they eat, the type of food they consume, and the exercise they choose to do.

I have added something new to the weight loss debate. From the results of some research I did some time ago, I have listed the foods that are likely to support our artery health and those that might cause it to deteriorate. I have used the results of my research into food nutrients and the 'furring' of arteries (the cause of angina, heart attacks, and some strokes) to direct the food choices of those trying to lose weight.

The weight we refer to as excessive, sits in the fatty tissues beneath our skin and in our abdomen. These surplus energy stores have some biological purpose, and an evolutionary advantage, but only in times of famine.

Body weight has now gained a political perspective. The growing prevalence of obesity is about to cost many nations a lot of money. Worldwide famine has yet to be eradicated, at the same time as many western countries flaunt their surfeit of food, available to all but a few.

No nation has yet addressed one important social topic successfully. That is the health divide, or the health inequality that exists between the rich and the poor.

There remains a big difference in nutrition, smoking, stress, body weight, morbidity, and mortality, between the socially advantaged and the disadvantaged. Obesity is more common among the socially disadvantaged, who also suffer three to five times more cancer and heart disease during their lifetime, than those who are better off and better educated. One important fact is that the disadvantaged cannot easily afford to buy protein. Instead, they buy much cheaper, fat and carbohydrate-rich food.

Because of its cost to health services, many western governments have decided to reduce the prevalence of obesity, to accept the inevitability of poverty and their inability to rectify it. For sound financial and politically astute reasons, some nations are focusing their attention on reducing obesity, rather than eliminating the poverty and the inadequate education that are its major contributors.

PART ONE

WEIGHT WORLDS

Chapter 1

Two Cultures: A 'Fat Mentality'. A 'Fit Mentality'

Some prefer to learn from a teacher. Others prefer to teach themselves. We can all learn from appropriate examples, with knowledge illustrated by metaphor, myth, parable or fable.

What follows illustrates the different features and outcomes of a 'Fat Mentality', and a 'Fit Mentality', and how each relates to self-gratification, physical activity, happiness, self-fulfilment and body weight.

The islands of Fitrakia and Zobesia, separated by only two miles, each had its own distinct culture despite a common ancestry.

In many parts of our world today, there are places where many proudly display the results of their 'Fit Mentality', while others are embarrassed by their 'Fat Mentality'. Not so on Zobesia. There, having a 'Fat Mentality' is applauded, and a source of pride. Zobesians think that being controlled by a 'Fit Mentality' is a waste of time and energy.

The cultural differences between the two islands resembled those of the ancient Athenians and Spartans (c. 500BC). Many Athenians worshipped the god Dionysus (Bacchus) and chose a life of decadence and indulgence. Spartans followed the god Apollo, and aspired to discipline and a frugal life dedicated to learning and rationality, fitness and warrior virtues.

Zobesians pursued ease and convenience, with lives dedicated to self-gratification, leisure and pleasure. As a race of well-rounded individuals, the prevailing shape of villagers helped to give them names: the Quadratis, the Spherims, and the Jabbarines (pear-shaped). They had mostly inherited genes known to promote weight gain. Their happy-go-lucky nature and relaxed demeanour, combined with a notable dislike of disharmony and stress. Self-gratification was their main source of happiness.

Zobesians adored luxury and dreamed of being engulfed by it. They demanded attentive service and convenience wherever they went. They kept their need for walking to a minimum, frequenting only those restaurants and shops that allowed their chauffeurs to drop them at the door. They abhorred shopping malls; too much walking was involved.

Zobesian homes had no staircases. Those houses with several floors had lifts. Those with a swimming pool and gardens, had them for aesthetic reasons only. They designed their living spaces for lounging and feasting, not for working or exercise.

Zobesians had no care to count the cost of their effortless, extravagant lives. They had oil, and it had made them rich.

All Zobesians employed servants from distant lands to provide help day and night. For this they paid very generously. House servants started their day preparing food. While at home, Zobesians insisted on having buffets displaying freshly prepared food at all times. The servants' next duty was to serve the family members as they awoke. They were required to wash and dress each of them. Washing and dressing was far too taxing for most Zobesians.

Capitalism thrived on oil-rich Zobesia, but only those business owners who pampered the self-indulgent whims of their customers, made money.

Zobesians judged one another by their trappings of opulence, and their body weight. It was an offence to criticise anyone for being 'overweight'. Zobesians were a race of large people, and being referred to as overweight in a negative or derogatory manner, was unacceptable. They had no wish to be compared to slim people, like the Fitrakians. In particular, they refused to accept what Fitrakians defined as obesity. They thought Fitrakians were underweight, and feared becoming like them. On Zobesia, the most notable members of society, carried the most weight. Zobesian society defined the 'Fat Mentality'.

On Fitrakia, life was austere and functional. Like the Amish in the USA, luxury and convenience were nowhere to be found. They ran, skated or cycled everywhere. There were few other vehicles. Home delivery as a convenience, was an unknown concept, contrary to their nature.

Fitrakians all had houses with a vegetable garden, and a place for a cow, goat or chickens. The interior style of their homes was unimaginative, white being the only colour used for their study, music room and gymnasium. Their kitchens were equipped for basic food preparation only. Pre-prepared foods were unheard of. No home had central heating. When heating was required, they chopped wood to burn.

No Fitrakian possessed more than two sets of plain clothes; one to be worn for work, while the other was being hand laundered. Because putting on weight was unacceptable, Ftrakians only needed one size of clothes.

The kingdom of Fitrakia was a commune, but with a limited form of capitalism. Their king never interfered (even though he had the power to do so). Each family was expected to be self-sufficient, except for the production of some basic manufactured items crafted by local engineers.

Sharing and helping others each day was one source of their self-esteem. Fitrakians never related worthiness to possessions. Only a fit body, and an enquiring mind counted. Fitrakians were a highly intelligent race who extolled the virtues of individual difference. The concept of equality in either the human or animal domain, was thought false and biologically unsustainable (evolution had been based on differences, not equality). They did, however, promote equal opportunities for learning and developing skills. Almost every citizen studied music, science and the arts (the most accomplished were known as Acadamenes). Most men admired physical fitness and fighting prowess, with only a few dedicating their lives to sedentary bureaucratic services (the Bureauniks).

Fitrakians despised ignoble ease, undeserved pleasure, and any form of self-indulgence. They ridiculed their Zobesian neighbours for being grossly overweight. Theirs was a fully functioning, 'Fit Mentality' society.

Although Fitrakia and Zobesia had existed for millennia, their existence remained undiscovered. Both islands enjoyed their 'unknown, unknown' status. (Many will remember U.S. Defence Secretary Donald Rumsfeld being mocked when in 2003, he tried to clarify a point about the War on Terror in the Middle East. His expression, 'unknown, unknowns', unfairly won him the annual 'Foot in Mouth' trophy).

Fitrakian minds were trained on reality, not fantasy. Their teachers and sages never encouraged speculation, except in academic discussion. They allowed no administrative and social guidelines; they thought guidelines too weak, and subject to too much interpretation. Their society had well defined rules for every action and response, including what to eat, when to eat, and how to exercise.

Fitrakian education was designed to be challenging. Most studied for university degrees, with many attaining higher degrees in their subject of interest. They encouraged those unable to attain good school grades to emigrate and become servants to Zobesian families. They paid them generously to emigrate.

The aim of the Fitrakian State was to create social stability through education. All citizens were encouraged to study philosophy and psychology, and to pursue artistic, scientific and intellectual pursuits that were altruistic, not commercial. Attaining fighting fitness was also encouraged.

Their ruled-based behaviour, blinkered attitudes, and intellectual pursuits, made them seem boring to many outsiders. Lean in mind and body, was how one Zobesian psychologist once described them.

Fitrakians did not believe in a life dedicated to trivial pursuits, self-gratification or vanity. They failed to understand why Zobesians were not bored to death with their lives. In private moments though, many would concede that Zobesians were happy: 'Fat mentalities in happy fat bodies' is how one Fitrakian psychologist described them.

The Zobesians gave no thought whatsoever to what Fitrakians thought. 'Live and let live' was one of their mottos.. 'Why don't they loosen their yoke of discipline, get a life, and enjoy themselves occasionally?' many asked.

Zobesian newspapers rarely mentioned politics or conflict. There was no need. They had a wise, fair-minded king in charge. He sorted out most of the problems without bothering his subjects. Any discussion of controversial subjects, they to left to the Fitrakians.

On Zobesia, news of the latest food trends, restaurants, health farms and relaxing spa resorts, filled their thoughts and newspapers. The effort required for study, for physical exercise, musical instrument practice and politics, they thought unnecessary. Their thinking was: why learn to play a musical instrument, when you can so easily hire a professional musician to play one for you? They regarded clever and talented people as useful, but not admirable. One of their favoured aphorisms was 'aspire to consumption, not to gumption'.

Comparative health issues had been studied, and were interesting. Although the average bodyweight of a Zobesian was two to three times that of the average Fitrakian, neither were particularly prone to illness. They had both inherited healthy genes from their ancestors. Most lived to 104-years of age.

On Fitrakia, suitability for immigration was decided after only after candidates took four tough, qualifying tests. The first test related to body size and shape. It simply involved squeezing through one of several small doorways (age-related), placed around the immigration hall. Being overweight (by their definition) instantly barred candidates from Fitrakian citizenship. The second test was to solve a series of difficult IQ puzzles. The third was to find the centre of a maze within a limited time, and then to build a shelter with the materials and equipment provided. (The aim was to exclude the unintelligent, and those lacking in practical skills). The fourth test involved assessing body strength and aerobic fitness in a gymnasium. To be successful, each candidate needed to pass each test. Every citizen under 80-years of age, had to pass the tests every year, in order to remain.

The Zobesian immigration tests were completely different. 'The Big Dipper' test was one, and the only other one was a measure of net financial worth.

Successful candidates were required to displace a certain volume of iced-cream enriched milkshake, from one of three large wooden vats. Each contained a different flavour of milk-shake, labelled, 'Chocolate', 'Vanilla' and 'Strawberry'.

After a shower and a change into the briefest of swimwear, each candidate would be immersed in the vat of their choice, and the volume of displaced milk-shake recorded. If the candidate's body volume complied with the Zobesian standard, they passed the test.

Only those with sufficient funds could qualify for Zobesian citizenship. The purchase of 100,000 shares in ZOC., the Zobesian Oil Company, was obligatory. Because of the high share price, only multimillionaire candidates with no debt could qualify.

Zobesian generosity, allowed all candidates who nearly passed the Big Dipper test, to try again. On offer was a two-month stay at an 'all-expenses-paid', 5-star luxury feeding station (an annex of The Gratifis Grand Hotel), before attempting the test again.

The Big Dipper test procedures, held every 3 months, were crowd-pulling events. Free food and drink were on offer to all onlookers. Large crowds of large people gathered to watch each event. It was a spectacle not to be missed; an enjoyable day out for the whole family. The desire to get something for nothing was a Zobesian trait, entirely consistent with their self-indulgent, 'Fat Mentality'.

Big Dipper spectators rarely left any food or drink. Those too satiated to move at the end of the day, were returned home by State transport. This thoughtful policy, typical of Zobesian generosity, supported their insistence on avoiding all effort.

For the citizens of both islands, it was strange that no gender or social equality issues had ever arisen. There were, however, major differences in lifestyle and culture between the two islands. How did these evolve? A little history will explain.

Back in the day when the athletic, serious minded, Prince Fitro came of age, his father allowed him to choose the lush mountainous island of Fitrakia to rule as king. He thus became King Fitro. He thought the terrain just right to physically challenge his citizens. His happy-go-lucky brother, once Prince Zobesi, now King Zobesi, was happy to take on the island of Zobesia as his kingdom. It was completely flat and oil rich. That suited his mentality. International oil sales meant that Zobesian citizens never had to work, and no citizen needed to be challenged by any unnecessary physical effort.

On Fitrakia, they only ate food as fuel, and only at set mealtimes. Fitrakians learned from a young age, just how much to eat in order to maintain an athletic body and a sharp mind. That meant eating only fish, shellfish, meat, fruit and some home-grown vegetables. The only time they ate carbohydrate, was before a sporting event.

Daily study led to a learned race of people, happy with a disciplined lifestyle. Their schools and institutes catered for the academically talented and gifted. Celebrity status came only to those who had gained university degrees, those who had published the most patents, and those who had won the most sporting medals. Others who were recognised for creating admired artworks, or publishing mind-changing books. Fame never came with wealth or possessions, and never because of personal beauty or charisma.

Zobesian culture had only loose guidelines:

- Do whatever it takes to save energy and avoid discontent.

- Because eating is one of life's great pleasures, eat as much food as you like.

- If there is nothing better to do, eat, drink and relax.

- Delegate every task possible.

In the royal households of both islands, life was not all plain sailing. On Fitrakia, King Fitro's son Prince Zagluton (who took his inheritance from his mother), desired a life of ease and self-indulgence. He became overweight by Fitrakian standards, and was not allowed to be seen in public. The thought of physical work made his legs feel weak, and he would have to lie down until the feeling wore off. It was no surprise when he failed the annual Fitrakian Citizenship Test, even though those who failed, risked exile (by mutual arrangement) to Zobesia.

Zagluton's failure to lose weight and change his ways, would normally have meant a spell at a military boot camp, eating minimal rations, and being worked hard physically. He was offered a few months stay at the austere Nonvuolamino Bay boot camp, 100 cubiti from the capital, but he rejected the offer.

Nothing in his genetic code, hormones status or brain chemistry, allowed Zagluton to know how much food was enough. His daily desire for self-gratification, gained from eating large amounts of chocolate and cake, meant he would not suffer the carbohydrate deprivation necessary to pass the citizenship body size test. Instead he elected to be deported to Zobesia.

Unlike his father, Zagluton had a richly developed 'Fat Mentality', with genes that promoted his appetite and weight gain. With some relief, and much sadness, his father King Fitro, agreed to his banishment. The old mutual, Inter-Island Agreement of 5700, allowed each island to banish their non-compliant subjects to the adjacent island.

So it was that Zagluton arrived on Zobesia. Here he came under the generous and supportive care of his like-minded uncle, King Zobeso. Although he missed his parents, nostalgia did not affect him. He knew, this was where he belonged. Crowds gathered to give him a rousing ovation, welcoming him to his natural clan.

Prince Zagluton, as you might have guessed, easily passed the Big Dipper test. He also had sufficient funds to purchase 100,000 shares in ZOC., the Zobesian Oil Company.

Prince Gymi of Zobesia (King Zobeso's son) also had his problems. He was as different from his father as a double chocolate ice-cream fudge sundae, is to a carrot! Self-indulgence was anathema to him.

Zobesians were disappointed in their Prince. From an early age, he would eat only protein, and eat only once every day when hungry.

He was underweight by Zobesian standards. Never was he seen to enjoy the buffets provided to passersby in every village. Was he unable to enjoy food? Was he ill? Why were his only interests studying, composing and playing music, and physical training? He had clearly inherited Fitrakian genes (from his mother and one grandparent). It came as no surprise, therefore, when someone at his father's court questioned his parentage.

From an early age, he had tried to hide his interests. As he grew older, he would exercise and study at night, thinking nobody would notice. He was discovered once too often, and when the reprimand came, it was swift and uncompromising. Banishment became the only option left to his distressed father. Any other action would have led to a public unrest. The day soon came when he had to take the boat to Fitrakia. There he would come under the disciplined eye, and strict tutelage, of his like-minded uncle, King Fitro.

It was strange, but true. Here were two cousins, better suited to the ways of their uncle. Their feelings of compatibility with their newfound culture, were immediate and cathartic. They had finally found the lifestyle that best suited their mentality. Relieved of cultural conflict, they found happiness, and the freedom to pursue their natural inclinations.

Chapter 2

Is Yours a 'Fit' or 'Fat Mentality'?

Losing weight is the subject of many books. Those who want to lose weight need to understand the role of what they eat and the exercise they do. Perhaps more important for them to ask is, 'Have I got a 'Fat Mentality', and 'Would I be more at home on Zobesia or Fitrakia?'

If you regard yourself as overweight, you might not believe that having a 'Fat Mentality' plays much part in your problem. Given that your well-intentioned parents encouraged to eat well as a child, your attitude towards food could seem normal and acceptable. If you are happy with your weight, and not locked into a vicious circle of repeated dieting, that is OK. If you have tried several diets in the past, but eventually lost motivation and returned to unrestricted eating habits, then perhaps you have a problem.

Next ask yourself, 'Am I more comfortable with fantasy than reality? If you are comfortable with fantasy, you are not alone. Many prefer to see themselves as potentially successful and famous: the next big sports star, better known than David Beckham, or the next

World Snooker Champion, more talented than Ronnie O'Sullivan. How famous would you like to be? Perhaps enough for mega-rich companies to offer you zillions, just to appear in one of their perfume adverts. If all you want is to lose weight successfully and maintain the loss, you must drop the fantasy of finding a wonder pill or diet (for the moment at least. Pharmaceutical companies are working hard to procure a cure for obesity). Changing your lifestyle and eating habits might seem unnatural, but both are essential for permanent success, even if aided by an appetite suppressant or the latest weight loss injection.

Although hope springs eternal, vain hope rarely wins battles. Facing up to reality, as painful as it can be, more often leads to success. If you are reading this to find an easy way to lose weight; a softly, softly, 'come on, you can do it' approach to weight loss, you have the wrong book. If you are reading this to find your next quick fix before lapsing back onto the same old vicious cycle of weight-gain and weight-loss, buy another book.

I don't want you to fail, but your 'Fat Mentality' (if you have a dominant one) will do everything it can to stop you succeeding. If you don't recognise the devious intent of your 'Fat Mentality', you might lose weight in the short term, but not in the long term. This is a simple lesson, but one that is hard to appreciate.

If you need help to recognise a 'Fat Mentality', ask yourself these questions next time you decide to eat:

- 'Am I actually hungry?'

- 'Do I really need to eat this?'

Next time you need to buy food and get into your car to travel a short distance to the shops, ask yourself:

- 'Am I being lazy?'

- 'Might it not be better for me to walk (given that I have the time)?'

If you think the correct answers to these four questions are 'No', 'No', and 'Yes', 'No', your 'Fat Mentality' is controlling you, and is suppressing your 'Fit Mentality'. Maybe you are genuinely hungry, and so short of time that driving to the shops is necessary. That is for you to question.

If you don't agree with any of this, and still want to lose weight, give it one last try, and read about Fitrakia and Zobesia again. If you still don't get it, eat, drink, and be merry. Be happy with the way you are. Stop trying to achieve the impossible. Get a large box of your favourite chocolates, put your feet up, start reading about the rich and famous, and enjoy life. The Duke of Wellington, Napoleon's adversary, said something like, 'During a battle, any idiot can stand fast and be cut to pieces. It takes a great commander to admit defeat and retreat.'

Let's assume that you have made some progress. You have recognised your 'Fat Mentality', and know when it is controlling you. If you have come this far, you have taken the first step to controlling it. If you have actually said 'No' to food occasionally, you have made progress towards losing weight. Well done, but don't get complacent. Your battle with a 'Fat Mentality' will never cease. Remember that your 'Fat Mentality' is ever-present, and will take every opportunity to gain control of your eating habits and physical activity.

With eating and exercise, we all harbour both a 'Fat' and a 'Fit Mentality'. They are constantly battling with one another, fighting to control our eating habits and exercise behaviour. If your 'Fit Mentality' loses too many battles with your 'Fat Mentality', chances are you will gain weight. Only by defeating your 'Fat Mentality' can you hope to achieve long-lasting weight loss.

To some extent, our lifestyle has been imposed upon us all from an early age. We cannot easily choose to lead a 'fit' or 'fat' lifestyle; our natural inclinations and circumstances lead us to one or the other. During life, we can flip between the two; alternating the time we spend with each mentality. Some will get stuck with one mentality, even when they have a strong desire to change. Many are happy to maintain a constant 'Fit Mentality', while others are happy with their 'Fat Mentality'.

Why might we get stuck with a 'Fat Mentality'? Vanity is one reason, conceit, complacency, depression, tiredness, laziness and a lack of discipline are others. Those aware of their age creeping upward, might choose to embrace their 'Fat Mentality' vigorously, rejecting all need for dieting and fitness. They may feel they cannot be bothered, and deserve to 'take it easy'. They may want to enjoy their riches and spend their savings. They may feel that they have done enough and deserve to be self-indulgent. This is understandable if we reflect on what use western societies make of those over sixty-five years of age.

The 'fit' and 'fat' mentalities affect our lives. Although they affect how healthy and active we feel, diseases are not so easily influenced. In large groups of people, physical activity, smoking, drinking alcohol, obesity, job status and poverty, can all be used to predict the number of heart attacks, strokes and cancers that will occur. It is not only obesity that counts. Our weight is only one of many powerful factors that can be used to predict morbidity (the diseases we suffer) and mortality.

It is in the nature of a 'Fat Mentality" to allow contented weight gain, and to act against successful weight loss. Those with a 'Fat Mentality' and a vain interest in fashion, may have to live a conflicted life, since the fashion world has long preferred skinny models. Should overweight people try to lose weight, and gain some self-esteem from attempting to be a model? Should they be content, whatever body form they see in their mirror?

To become overweight and get rich, both require a similar outlook. A 'Fat Mentality' will be of help to both. Both must want to gain reserves in excess of need.

If you are a business 'fat cat', you will have become so either by being given money (family inheritance, a lottery win), or by making money (doing deals and running a business). It is almost impossible to become rich by selling your time and labour. From the start, 'fat cats' aim to gain more money than they need. Some have an exaggerated need for financial security, and come to regard one billion GB pounds or US dollars, as not enough. For many overweight people there is a parallel. They will have accumulated a wealth of energy, stored as body fat, which is much more than they need for everyday living.

The financial 'fat cat' stores his / her excess reserves in a bank, as bonds, stocks, shares or property (whichever gives the best unearned return). A 'Fat Mentality' demands the highest rate of interest for the least amount of work. The overweight person stores 85% of their excess energy under their skin and in abdominal fat. Their extra fat reserves put them in a state of energy surplus.

The rich and the obese have something else in common. When a business 'fat cat' accumulates more money than he, his family, or his progeny will need, and when an overweight person stores more energy than they need to survive months of famine, neither will find it easy to stop accumulating more. Why is this? Is it greed, or is it their inner nature that drives their compulsion?

My friend Gerry accumulated hundreds of millions of pounds by the time he died in a tragic accident. He once told me he had grown up in Ireland without shoes on his feet. He had taken a vow never to experience poverty again.

If you eat more than you need and just cannot stop doing it, even though you are unhappy with your weight, your 'Fat Mentality' is firmly controlling you.

If you are overweight and wish to slim down, you will need to understand how to control your 'Fat Mentality'.

If you lose some weight without attending to your 'Fat Mentality', you stand little chance of maintaining the loss.

Everywhere in the western world, food is available on most street corners. That doesn't mean we should blame food retailers and chefs for our obesity. If you are prepared to admit the following, you may be ready to lose weight: 'My attitude and outlook to life (my mentality) is as big a problem as my eating and exercise.'

Although our body weight and shape is mostly a matter of personal responsibility, many try to transfer some blame for it to their diet or to others. They might blame their personal trainer, or their nutritionist for misinformation. They might blame their friends and family for not supporting or encouraging them enough.

Our Mentality Defines Us

To lose weight you are going to have to accept total responsibility for your eating and exercise habits. Your 'Fat Mentality' will drive you to take it easy, even when there is work to be done. It will drive you to eat for pleasure, even when you are not hungry. It will encourage you to take the easy path at all times; to drive instead of walk; to order a takeaway, rather than to cook for yourself.

Your 'Fit Mentality' will drive you to walk rather than ride, to dismiss thoughts of food until you are genuinely hungry, and to challenge yourself to overcome adversity. Only after controlling your 'Fat Mentality', and promoting your 'Fit Mentality', should you ask, 'Which diet regime should I follow?'; 'What should I eat, and how much?', and 'What exercise should I consider?'

Knowing what to eat in order to lose weight, and possibly prevent heart disease, requires some specific knowledge about food and lifestyle. That's the easy part!

If you are still not sure about the consequences of your opposing 'fit' and 'fat' mentalities, take the simple questionnaire test in the next chapter.

Chapter 3

What Mentality Controls You? Take the Test

Inside our heads, we continually talk to ourselves. This is our inner voice. It is how we think and provide the subtitles to our emotion dramas. Our thinking can be fast, reflexive and instinctive, like deciding to buy something on the spur of the moment, or slow and considered, like deciding the details of the table plan for a wedding.

In his landmark book, Thinking Fast and Slow (2011), the Nobel Prize winning psychologist Daniel Kahnman, called these types of thinking, Type 1 & 2. Both are active as you walk around a shop, deciding which foods to buy. Both are active as you choose food from a buffet. But what shapes our thinking in the first place?

At every food choice moment, it is your dominant mentality ('Fat' or 'Fit'), that will direct your thinking and guide you to your final choice. The winner of this contest will vary from time to time. You can be ambivalent, and not sure of what to choose, but mostly one mentality will dominate. The dominant mentality gets to control your immediate food choices, and later on, your body weight, physical fitness and body shape.

If you are overweight and unhappy about it; if you have failed to lose weight or have failed to maintain weight loss, you are most likely under the control of a 'Fat Mentality'.

Not only the overweight eat a lot; slim, fit athletes do also. They eat energy dense food and only rarely get fat. Some overweight people are quietly in awe of them. You will hear them say, 'They're so lucky. They can eat anything, and not put on an ounce.' Do you find this mystifying? If so, you are not alone, since many nutritional scientists have yet to fully explain it.

Whenever we try to classify any biological phenomenon as black or white, fit or fat, there will always be some exceptions. No human is forever one thing or the other. Our nature is plastic and changeable. Therefore, no law or set of rules made to control or define us, will ever apply at all times.

Have you ever seen a professional athlete eat? I saw one once in a steakhouse restaurant owned by friends of mine (The Phoenix Apollo in Stratford, East London). He ate a mountain of pasta, followed by the largest T-bone steak I have ever seen. He ate like a glutton (one manifestation of a 'Fat Mentality'), yet in every other way, his 'Fit Mentality' was in charge.

Ask yourself this question:

After enjoying one of six doughnuts in a pack, can you resist eating another?

If 'No', would your reasoning include any of the following?

'I like them a lot, so why shouldn't I eat them?'

'Since I've just come back from the gym, I can afford to eat more,'

'I always eat what I want!'

'I work hard, therefore I deserve to eat what I like!'

'Life is too short, so I want to indulge myself!'

'I don't care anymore!'

These are the common responses given by those who fail to lose weight or fail to maintain their weight loss. They have resigned themselves to their 'Fat Mentality'.

Adaptation

It may be difficult to exorcise a 'Fat Mentality' if it has been held strongly over a long time. Its dominance over a 'Fit Mentality' might become accepted and never questioned again.

Let's take a parallel example. A young man with widespread facial acne, is seriously embarrassed by his appearance. He might stop going to social gatherings, believing that everyone is looking at him and discussing his disfigurement. Because he is self-conscious, he might assign every laugh and snigger, to his appearance. It wouldn't be, of course, but that is how he might think.

He might gradually learn to be less sensitive and to adjust. Slowly, he might come to accept acne as part of his life. Having found no doctor who could help him, he might resign himself to accept his appearance and ignore it.

This adaptive process can happen to any shy person with what others judge to be a 'disfiguring' medical condition. It happens to many who have been overweight for a long time. They may find it embarrassing, but become resigned to it.

After years of being told to eat less and exercise more, some overweight people will become disinterested and disheartened. Denial may be unrealistic, but it is a powerful way to cope with problems. Becoming insensitive will protect us from concern and embarrassment, but will inhibit us making beneficial changes. For those with little motivation to help themselves (if disinterested and disheartened), losing weight may no longer be seen as a plausible option.

The problem doesn't end there. Those who have lost weight successfully, might not like what they see in their mirror; some might miss the psychologically adapted, disinterested mind-set, they once enjoyed when they were overweight. For some it will be easier to maintain an overweight state than a slim state, given the vigilance and self-discipline required. Give up on vigilance, and weight gain is inevitable. Although it may perplex others, there is nothing actually wrong with dismissing food vigilance, being disinterested in your weight problem, and accepting the way you are. At least, you would qualify as an ideal Zobesian citizen.

It may define you as having an entrenched 'Fat Mentality', but who other than you, really cares?

So What is Your Dominant Mentality?

Complete the following questionnaire, to find out.

- Each question has one of four answers.
- There are no 'right' or 'wrong' answers.
- Try to be completely honest with yourself.
- We are all capable of fooling ourselves, so try to get in touch with your true self.
- Ring the answer closest to how you would normally react.

Question 1: You have eaten recently. You are offered a plate of food you really like. Do you:

A: Refuse it, and mean it. **B:** Refuse it, but don't actually want to.
C: Accept it, but eat as little as possible. **D:** Eat all of it without a second thought.

Ring your answer: A   B   C   D

Question 2: You are home alone. You want a cup of tea. You love doughnuts. You are trying to lose weight, but faced with a pack of six jam-oozing doughnuts, do you:

A: Refuse to eat any, but then give in. **B:** Eat the lot without a second thought!

C: Eat only one half of one doughnut. **D:** Find it easy to eat none of them.

Ring your answer: A   B   C   D ?

Question 3: Your guests arrive in two hours. You are out of coffee. The nearest shop is a 10-minute walk away. Do you?

A: Walk because you enjoy it. **B:** Send someone to get it.

C: Walk because it is good for your health. **D:** Jump in the car. It's quicker.

Ring your answer: A   B   C   D

Scoring the Test

Let's see how you scored. The maximum score for each mentality is six.

Your 'Fat Mentality' Score:

Question 1: If you ringed B score '1'.
If you ringed D score '2'.

Question 2: If you ringed A score '1'.
If you ringed B score '2'.

Question 3: If you ringed D score '1'.
If you ringed B score '2'.

WHAT IS YOUR TOTAL 'Fat Mentality' SCORE:............

Your 'Fit Mentality' Score

Question 1: If you ringed A score '2'.
If you ringed C score '1'.

Question 2: If you ringed C score '1'.
If you ringed D score '2'.

Question 3: If you ringed C score '1'.
If you ringed A score '2'.

WHAT IS YOUR TOTAL 'Fit Mentality' SCORE:............

What do your ''Fat & Fit Mentality' scores imply?

A Bit of Background

We all have a 'Fat' and a 'Fit Mentality', but we differ in how much we allow them to control our eating. We also differ in how long we allow each to stay in control.

Our eating and exercise behaviour can sometimes be a source of guilt. Many see overeating as letting themselves down. Their self-esteem reaches a low point, their 'Fit Mentality' retreats, and they resign themselves to be controlled by their 'Fat Mentality'.

Our dominant food and exercise-related mentality influences our self-control, or lack of it. Our mentality influences what we choose to eat, and what we choose to do from one moment to the next. Both fast and slow thinking are affected.

Like genes, our mentality can exert a strong or weak influence. Some people have strong forms of the 'Fat' and 'Fit Mentality'. Addictive personalities, for instance, may have such a strong tendency for self-gratification, it gets beyond their control. They may have an inability to say 'No' to any habit, whether it be eating, smoking, drinking alcohol or addictive substance abuse. These are aspects of a strong 'Fat Mentality'. Others with a strong 'Fit Mentality' will not reject the hardship of lengthy periods of fasting, frequent exercise or self-deprivation (as illustrated in the film, 'Forest Gump'). The more perfectionist, nit-picking and obsessive the character, the more likely they are to have this as their dominant mentality.

Technical Point: A wide spectrum of mentality types (biological diversity) has allowed selective survival. Variation is required to allow only those best suited to their environment to survive healthily. In our chosen environment, our dominant mentality will predict how fit we are for healthy survival.

The questionnaire I designed to test food and exercise-related mentality is basic, and open to much error (misinterpretation and variations in how we answer). Answered truthfully, it will at least detect the extremes of our mentality.

Your Total 'Fat Mentality' Score

If you scored three or less, you should be able to lose weight without trouble.

If you scored between 4 and 6 for 'Fat Mentality', you have a significant mentality problem, and may find it difficult to lose weight. You can change, but you may need to take heed of all that follows if you are to suppress your 'Fat Mentality' most of the time. You may need to consider regular counselling, psychotherapy (cognitive therapy) or hypnosis, to help you better understand why you eat what you do, why you eat as much as you do, and why you choose to exercise as you do.

If you scored between 4 and 6 for a 'Fit Mentality', you have no significant mentality problem in relation to your weight. You should find it easy to lose weight by eating less and exercising more.

So You Have a 'Fat Mentality'

By now you should know what I mean by a 'Fat Mentality'. There are, however, different aspects to it, other than eating, getting fat and becoming unfit.

Let's look at how this mentality affects money and activity.

Here is a list of 'Fat Mentality' Characteristics:

Applied to Food:

- Food easily seduces you; so does the food being eaten by others.
- You are happy to finish the meals of others.

- You often eat when you are not hungry.
- Despite not being hungry, you will eat at the suggestion of others.
- While eating, you prefer to keep your mouth full all the times. You keep food on your fork

  in readiness.
- Grazing is your style of eating.
- You may see large portions as insufficient.
- You have reverence for food, drink and restaurants.
- You prefer energy dense, fatty and starchy food.
- You use food as a reward (for doing exercise, and for 'being good' when keeping to your   diet).
- As a host or hostess, you are generous with food.
- You may prefer 'all-inclusive', pre-paid buffets, with 'eat as much as you like', and 'Go Large' options.

Applied to Activity

- You love the easy life and always choose the easiest paths through life.
- You have no special inclination to exercise either your body or you mind.
- You revere of all forms of comfort and luxury.
- The prospect of pleasure and leisure easily seduces you.
- You measure success in terms of excess.

Applied to Money:

You may think:

- A 'good business' is one that makes a large profit with minimal effort.
- Wealth improves your security and allows you to take it easy.
- Delegation is a crucial strategy for you. It produces more income for less personal involvement in work.
- A continually accumulating fund is essential, even if you are already wealthy.
- You need to escape poverty.
- Large amounts of money may seem small to you.
- Rich people are to be revered.
- Philanthropy is a good thing, especially if tax deductible.
- Feeling rich is good.
- Shopping and spending money gives you a buzz. It can become an essential fix (1st class travel, and the best of everything, regardless of your actual wealth).

If you still want to lose weight, you must recognise which mentality controls you: a 'Fat Mentality or a 'Fit Mentality'. You will not find it easy to lose weight, if a 'Fat Mentality controls you. Only after acknowledging this can you hope to conquer and control those behaviour patterns and attitudes that caused you to gain weight in the first place.

If you are not yet ready to face yourself in a mirror, you may not be ready to continue. You may need to start this book again. If you cannot accept the truth about your mentality, and its command over you, revisit Fitrakia and Zobesia in Chapter 1.

PART TWO

DO YOU HAVE A WEIGHT PROBLEM?

Chapter 4

What Weight Problem?

Many want to alter their shape and weight, but do not know where to start, especially if they have battled many times with weight loss. What are they to do, when no diet plan works? Perhaps they have a hormone problem. Perhaps success requires a missing ingredient or secret formula. Perhaps an operation or a magic pill would work. This book contains some answers.

A life-changing moment for me came after I awoke one morning and looked at myself in a mirror! No cursory glance this time. What struck me was how much I had changed. What I saw, compelled me to do something about it.

I had lost some weight before on an Atkin's, low-carb diet, so I did the same again. After losing some weight, I intended to regain my physical fitness.

My first aim was to lose enough weight to feel comfortable. I wanted to feel satisfied with my reflection in mirrors. I wanted my fitness level to go beyond my everyday needs. I wanted to have more energy, and no obvious exercise limitations.

The more I thought about my clinical experiences as a doctor treating overweight cardiac patients, the more something struck me. Successful weight loss depends not only on understanding calories (food and exercise) and carbs, but on the acknowledgment of what drives or inhibits our behaviour - our mentality, shaped by all our cultural influences, by our personality, and by the personal attitudes we hold.

The lucky few have no problem at all with their weight. This book is not for them. Others find weight loss an unceasing battle. Many try repeatedly to lose weight, and fail repeatedly. I have written this book to help them.

Some of them have had to:

- Change their choice of clothes,

- Change their inclination to exercise.

- Change their sexual expectations.

- Change how they collaborate with others.

Is it their fault? Are they doing something wrong, or could their metabolism or hormones, be at fault? These are vital questions to be answered in view of the latest medical research.

The bullying of overweight people is not uncommon. As a result, some may become embarrassed and made to feel guilty about their size and shape. The assumption may be that gluttony and indolence are at work. Being called obese, podgy, well-rounded, plump, chubby, buxom, fat, well rounded or a roly-poly, can have deleterious effects on their mental health. In his book *My Amy*, Tyler James says the media called Amy Winehouse 'curvy'. She interpreted this as 'fat'. Apart from being constantly pursued by the media, this description troubled her considerably.

All those seeking a relationship are acutely aware of how much attractiveness counts when competing for a partner. Not everyone finds overweight people attractive. Some overweight people who care about this, will diet and exercise beyond reasonable limits. The motivation to change shape, and more easily find a partner (so they think), can be strong. Behind all this lies an evolutionary demand to survive (the survival of the fittest), and a deeply seated biological drive to procreate.

Are the obese a burden on society, given all the costly extra healthcare they are likely to need? In a democracy supporting equality, surely this is misguided thinking and bad behavior. It lacks respect for the right we all have to make our own choices, even if experts think them inadvisable. Overweight people have the right to reject sermons about the generalised risks of weight gain (that might not actually apply to them), and reject the notion of national health benefits and the money to be saved, should most overweight people lose weight.

There is a sound reason for upholding the right of every overweight person to remain as they are. The reason is, no doctor can guarantee a health benefit to any individual who loses weight, even though the statistics strongly suggest that the medical risk of group obesity, is many times higher than that seen in a group of ideal weight people (comparing one large group to another).

Changes in accepted fashion have always made it difficult for some overweight people to be content with what they see in their mirror. This negative view is now being magnified by public health medical professionals who issue regular risk warnings about the dangers of obesity, suggesting impending heart attacks, strokes and diabetes. A growing number of overweight people, have become tough-minded enough to reject such dire generalised predictions that might not actually apply to them as individuals. They would be right to question the relevance of group statistics applied to any individual, but is this a reasonable position, given that the statistics we now have are a reliable guide? Sensible or not, the overweight have the right to decide.

Those who decide to lose weight give many reasons for doing so. Some are real, and some are lightly sprinkled with fairy dust. Among them are a desire -

- To be fashionable.

- To get back into those old clothes (back to a younger self).

- To look younger.

- To feel more comfortable.

- To be less breathless.

- To look good in the mirror.

- To improve health and health risk.

- To improve self-esteem.

- To stop the bullying.

- To reduce the arm twisting of well-wishers, including medical professionals.

To successfully lose weight and maintain the loss, I have addressed eight topics, that follow on from the understanding and control of a 'Fat Mentality'. They are:

1. What is overweight and underweight, and what are the causes?

2. Appetite and satiation. Metabolism and hormones.

3. Food. What to buy and what to eat?

4. Portion sizes. How much food should we eat and when?

5. Exercise. If, what and when? How much, and how often?

6. Self-help.

7. Medical help: drugs and psychological services.

8. Surgical help: the various operations.

One of my aims has been to provide weight loss advice that is good for the heart and circulation, the health of which largely determines our life expectancy (except for cancer). The Atkin's method of weight loss, provides a diet that works, but can contain too much animal fat. Eaten long term, that might not favour artery and heart health.

Losing weight is one thing, maintaining it is another. Dieting alone doesn't work well if sustained weight loss is the aim. Dieting is unnatural for many, and can be an unhealthy pursuit for some. Despite this, millions wait for the next magic wonder diet; the next food fad and the next formula, pill or injection that will help them lose weight effortlessly.

If you are with me so far, indulge me a little further. Find a full-length mirror. Now look at yourself in the mirror. Are you happy with what you see (your shape and weight)? If you are, stop reading this book. Give it to someone who is unhappy with what they see in their mirror.

If you are reading on, ask yourself a crucial question. 'Why do I care about my weight?' It is important to acknowledge what matters to you, not what matters to others. NEVER allow your self-esteem to depend only on the value others grant you, however gratified you may feel.

If you are at one with this, ask yourself again,

> 'Do I really want to lose weight?'

From the Pulpit

To lose weight, you must have the discipline and the motivation for it. You may need to put your heart and soul into it. For that reason you may need a change of heart. A similar commitment is required to stop smoking successfully, to get qualified in a difficult subject, and to get fit. There is no way to soften the hardship involved in the weight loss process, but I hope to make it easier.

For the moment, those who are searching for a magic bullet pill or diet, will need to think again. Those who place self-gratification above any need for effort, are likely to find the hardship of changing habits unacceptable. Without a change of attitude, they might lose a little weight, but will it last?

Help From a Wizard Perhaps?

Now for a bit of New Age psychobabble. As Dorothy found (in the film *The Wizard of Oz*. MGM. 1939), getting back home was a lot easier than she thought; she had no need for magic. She survived a perilous journey to the land of Oz, to ask the Wizard how to get back home to Kansas. He didn't know. He was a charlatan. She was then told by a benevolent witch that she had always known how. Instead of asking gurus and wizards for a secret way, all she needed to do was look within.

When the going gets tough, it often helps to have some encouragement and camaraderie along the way. Like Dorothy walking along the yellow brick road, progress is made easier with the help of friends. A talking lion who needed to find his courage, a thoughtful tin men in need of a heart, and a scarecrow wanting a brain, are not easily found companions these days.

Are you a Denier?

- Do you avoid seeing yourself naked in the mirror?

- Have you spent lots of money on clothes to camouflage those excessive curves?

- Are you uncomfortable talking to others about food and weight issues?

- Do you wish you had a magic wand; one that could make your weight problem disappear?

- Do you avoid talking about weight problems, and being lectured by others?

If any of these apply to you, you may have put weight issues out of your mind, perhaps to deal with them at some later date. That could mean you are in denial (a problem seen as too difficult to confront), or perhaps you are not that unhappy with the way you are.

Do you avoid looking at yourself in mirrors? Do you avoid looking at yourself naked in front of a full-length mirror because you find your body somehow unacceptable? Burying your head in the sand is self-imposed ignorance, or denial that solves nothing. Both can relieve stress for a while. The pain of reality is stressful for those who find it difficult to cope with life. Putting off a problem until you feel ready to deal with it, can help to reduce stress in the short term.

Should you ignore the advice given to you by your best friend or doctor? One reason you might legitimately do so, is that your opinion counts more. Unless the need to lose weight is a strong conviction, you will waste your time, and that of any others who try to help you. There is, however, something worse than wasting time: risking further personal failure and losing self-esteem. This is not a good thing for anyone already unhappy, guilty or disappointed, with their past attempts to lose weight.

I was once listening to the radio when a female guest TV presenter with a weight problem said, 'Obese people are never happy. If they think they are happy, they are fooling themselves!' Was she exaggerating for the sake of effect? In my experience, it would be nonsense to suggest that all overweight people are unhappy or depressed.

There are many who are truly happy with their weight. They should forget any attempt to change it. They have the right to eat, drink and be happy. With excessive weight gain, you may (on average) risk living a shorter life, but while you are alive, you may well be happier than those who choose to stress themselves with weight loss regimes. Weight may put your health at risk, but then being miserable and unhappy can prove fatal. Doctors have for centuries, sidelined the health risks posed to life by failing mental health. Up to know, we have put most of our efforts into dealing with organic disease.

You need to know whether you really want to lose weight, and whether you have what it takes to do so. You might otherwise join the ranks of millions of dieters, many of whom have repeatedly failed to achieve successful, long-term weight loss.

If you are sure that you want to lose weight, repeat the mirror test to confirm it.

The Mirror Test

- Go to your bedroom, bathroom or hallway – somewhere there is a full-length mirror.

- Strip off and do a slow twirl.

- Do you like what you see?

- Are you happy with how you look?

- Let yourself go. Relax and let your stomach hang out all the way! See yourself at your worst.

- Now, repeat the slow twirl. Are you still happy with what you see?

Take your time and really look at yourself. You will now have the answer to the question:

Do I really want to change shape and lose weight?

If your answer is 'No', the only benefit you will get from reading this book will be to find out which foods are good for your arteries and heart.

Chapter 5

Who Said You Are Overweight?

Are you overweight?

Have you calculated your body mass index (BMI. See explanation later) or is it based on opinion? Is it you who thinks this, or someone else? If you are unhappy with your larger clothes' sizes, and unhappy with what you see in a mirror, it is possibly true. Occasionally, doctors see those with a distorted idea of their ideal weight and clothes' size. Some believe they are overweight when their BMI indicates they are not.

Your friends and family may think you are overweight. They may all have subjective reasons for their judgement, but should you agree with them? If you are not sure, refer to your body mass index (BMI). See later in this chapter, and Appendix 1.

There are several ways to judge your body weight and its relevance to your health, future health, and the way you want to lead your life.

The first thing to know is that you can be healthy, whatever your weight. This little respected fact will not suit those who believe that there is a close relationship between health and body weight.

There is a paradox here. There is a relationship between weight and illness, but only for large groups of people. What applies to a group may not apply to an individual. This little annoying statistical paradox; this little devil in the detail, should concern everyone who is told, 'You are overweight', 'You must lose weight', 'To be healthy, you need to weigh . . .'

Population statistics can give us reliable body weight averages, but why should you compare yourself to a depersonalised group average, or to others who may not share your personal features?

The medical profession now categorises body weight as ideal, overweight, obese and beyond. There are several ways of doing it: your clothing size, bodily characteristics, and using height and weight calculations. Before you accept the medical implications of obesity, you should know how doctors can use your height and weight to calculate your body mass index (BMI), and estimate your risk of death and disease. They will not, of course, be so easily able to estimate how content you are with your weight, how happy you are with your life, and how likely you are to realise your weight loss expectations. Any doctor trying to make such assessments will have to use an older form of medicine, now fast disappearing: the art of medicine.

Clothes Size

One easy way to judge bodyweight and shape is clothes' size. Slim men, bordering on being underweight, will fit into the smallest of shirt sizes (S), whereas large men could need XXXL sizes. Small women need dress sizes between 6 – 8 (27 inch waist). The average woman in the UK in 2017, needed size 16 clothes, with a waist measuring 34 inches (86.4cm); for larger women, dress sizes of 18 are equivalent to a waist size greater than 36 inches (91.4cm). Larger women need clothes of size 18 and over.

Waistline (abdominal girth) is a simple measure. It is a measure of abdominal fat, and of fat stored elsewhere. Some researchers have found the incidence of coronary artery disease, correlated to abdominal girth, although it is less reliable when fat is widely distributed beneath the skin. Despite a large abdominal girth, a sizeable minority remain free from chronic disease. The term 'metabolically healthy obesity' has been used to name it.

When a condition is exceptional, it is usually given a special name, like 'metabolically healthy obesity'. The problem is that the characteristics of human biology are always varied, making the use of averages of much less value. It could be said that each of us is exceptional.

There are other simple ways to classify body weight. Can you pinch an inch of fatty flesh around your tummy, or on the back of your upper arms? If you can, you are likely to be overweight.

Observe the effect of someone walking. As they walk, some large people sway from side to side. Their flesh may ripple or wobble as they do so. Small amounts of exercise can make them short of breath. Excessive levels of obesity have been found associated with dire medical consequences. Groups of people who are very large (see later: those with a BMI greater than 40), can have a three to five times greater risk of dying prematurely, than groups whose weight is measured as ideal.

You will have noticed on visits to a zoo that walruses and seals have flesh that ripples when they move. This is because they need a thick layer of fat to insulate them against Arctic and Antarctic water temperatures.

If we humans accumulate too much body fat, we too will vibrate, oscillate, ripple or wobble as we move. In some cultures, obesity signifies wealth; being underweight signifies poverty. In western cultures, being overweight is viewed as a sign of gluttony, over-indulgence or laziness. These biased judgements are disrespectful, and often unfair and incorrect.

Is wobbling flesh attractive? It is to some. Love handles and groin creases are features. When we try to wear old jeans and trousers (what my mother used to call 'slacks'), an awareness of how large our tummy and buttocks have become, will be obvious.

While walking in a swimming costume, any wobbling, rippling, oscillating or vibrating flesh will be noticeable. The younger the person, the greater the weight gain needed. Young people have firmer flesh than older people, although even young flesh is not always firm in excess!

As we put on weight, all our fat stores grow, both under our skin and in our abdomen. Some people specifically add weight to their face, thighs, bottom or belly, while others grow in all directions at once. When rolls of fat form under the skin, creases appear. The older we get, the more lax our skin becomes and the more readily it creases.

Having put on more weight than I realised, I once accused my optician of selling me spectacle frames that were too small. In fact, my face had grown fatter and my spectacle frames had become tighter as a result.

Now ask yourself:

- Do I wobble when I walk?
- Does any part of my body wobble when I run?
- Does any part of my body ripple when I clap my hands?
- Does any part of my body vibrate when I flick it with a finger?

We all have some fat under our skin. It insulates us from external temperature changes and protects our inner organs from the sun's radiation. In our tummies, fat forms drapes, hanging like curtains from our intestines (omentum). It quietly awaits the call for energy during times of famine and fast. The release of energy from our stored reserves then begins.

Several factors determine the storage of fat in the abdomen (central obesity). It is commoner in males (commonly referred to as a 'beer belly'). It is here that some store the excess energy found in beer and energy-rich foods.

A Little Technical Information: An increase in the all-cause death rate in the UK was found among men with a waist measurement of 95 cm (37.4 inches) and over, and for women with a waist measurement of over 80 cm (31.4 inches or dress size 14) and over. (see Carmienke, S. et al. 2013, in Bibliography). Other research has shown that larger waist circumferences are associated with an increased risk of coronary heart disease. In some groups, a 37% increase in risk has been found (see Hong, Y. et al. 2007).

In the fashion industry, being underweight has remained de riguer since the Second World War. For fear of gaining weight, many models will push peas around their plate, rather than eat them. Some who actually eat them will vomit them afterwards (bulimia). This is how some preserve their sylph-like form.

Body Weight Related Questions

Many of my overweight patients asked:

- 'Am I really that much overweight?'
- 'What weight do you think I should be?'
- 'Do I really need to lose weight?'
- 'Should I lose weight?'

You can check the published data for your supposed 'ideal weight' (See Appendix 1).

I don't much like these tables of data. The sheer weight of numbers implies accuracy, authority, reliability and applicability. They are based on population averages, not on what is best for you and me. They serve as a guide to that illusory number - the 'ideal weight'. Don't be fooled. There is no such thing as an ideal weight for any individual; there are just too many variables and too many other factors (metadata like racial origin, inheritance and lifestyle) to account for. The concept of an ideal weight provides a useful rough guide, but it is loaded with presumption. One presumption is that your ideal weight is 'best for you'.

A Few Key Questions

- If you are the shape you want to be, why worry about your weight?
- Are you happy with the body shape you see in a full-length mirror?

There are those of average weight, and others who are actually thin (perhaps anorexic or bulimic), who feel overweight, and see themselves as overweight in a mirror. Some of them attend weight reduction clinics with a distorted idea of their body shape and weight.

For most of us, the mere feeling of being overweight, is enough to motivate a dieting regime.

When I was considering losing weight, my concerns were:

- Whether I would ever get back into my old clothes (comfortably).
- Without weight loss, might I develop rolls of fat and creases in my skin?
- Without weight loss, would I wobble as I moved? For women, breasts of any size move in response to movement, but should men experience the same?

When naked, what did I see in a full-length mirror? Not a shape I liked, that's for sure! What I wanted back was my younger body shape; the one I remembered from a time when I was fit. What I saw in my mirror provided me with enough motivation to change.

Rather than focus just on weight and shape, think of the effects of being overweight. For instance, are you breathless,

- When you walk?
- When you try to walk and talk?
- When you bend down to put on shoes?
- As you dress?

If your answer is 'yes' to any of these, you have a weight related physiological problem. What do I mean by that? Clearly, your weight is affecting the efficiency of your lungs (because of diaphragm restriction). This will cause you to breathe more deeply and perhaps faster, in order to acquire extra oxygen you need.

An Important Point

Every half stone (7 pounds or 3 kilos) of extra weight you put on, will usually make you a little more breathless. Lose the same amount of weight and you will notice improved breathing.

WARNING. There are many other reasons for breathlessness. You could have a problem with your chest (lungs), your heart or anaemia. For those under 35-years of age, however, the odds are that weight gain alone is the cause. If you are over 35-years of age, and have heart disease in your family, think again. Seek a diagnosis from your doctor. For more detail on this subject get my book How to become Heart-Smart. A User's Guide to Heart Health and Heart Disease Prevention (2nd edition, 2024).

Body Size. Is it an Optical Illusion?

Have you noticed that a half-pint bottle never looks half the size of a pint bottle? It can look more like ¾ of the size. The reason for this is that we cannot judge volume well. This is bad news for those who want to change shape. They could lose a lot of body weight and still not look much different. The same applies to putting on weight. If you double your weight, you will not look double the size. To change your size a little (decrease your body volume), you must lose a lot of weight.

Overweight and Fit

Can you be fit and overweight? Can you carry excess fat and exercise without a problem? It is possible because a healthy body will adapt to the demands put upon it. An overweight body will make adaptation more difficult. For some sports, extra fat helps. At least long distance sea swimmers and Sumo wrestlers have an advantage when they carry excess body fat. I doubt that an underweight swimmer would survive long swimming the English Channel in winter.

Kim's Case

My patient Kim was a rather special lady. She was gracious and charming, and an ex-nurse of very generous proportions. She came to me complaining of breathlessness while walking just a few paces. Her lips also turned blue as she walked!

To my surprise, all her heart tests were normal. Her heart, which I had tested many times with ultrasound, appeared to contract normally and was of normal size. It had remained so, despite her long-term weight gain of 150 pounds. I concluded that her problem was equivalent to trying to power a battle tank with a lawn-mower engine. Her heart was not capable of circulating enough blood and oxygen during exercise. This caused her to become breathless with minimal exertion. Because her heart was too small for the job, it failed to pump enough blood to her lungs. She became blue on exercise (cyanosis), because she was unable to absorb enough oxygen.

A decade later, Kim developed angina, and I then confirmed her coronary artery disease. She died a few years later from a sudden heart attack, despite all the preventative drugs she took. Finding it too taxing, she never did lose weight.

Before blaming her weight for her demise, it is important to state that many patients with coronary disease are slim and physically fit. It is not only the obese who die with heart attacks. The causative artery 'furring' process (atherosclerosis) occurs regardless of body weight, and total blood cholesterol (the exception is HDL cholesterol which provides protection). To learn more about this important subject read my book: How to Become Heart-Smart. A User's Guide to Heart Health and Heart Disease Prevention. (2nd edition 2024).

Body Form. What Shape Are You?

After moderate weight gain, some women will develop larger breasts and buttocks. They actually grow no more breast tissue, only more fatty tissue. Men develop 'breasts' for the same reason. If you belong to this group, you will wobble when you walk.

Those who are grossly overweight, walk in a certain way, swaying from side to side. This is for several reasons. First they must pass one large thigh in front of the other while walking. In addition, large changes in their centre of gravity need to take place.

There are several overweight body shapes: the big belly (common in men), the big bottom, and simply big all over. Some become spherical, others grow to resemble a cube.

From what I have seen in swimming pools, observing mothers and fathers with their children, the inheritance of body shape is inevitable. The size to which they grow is another matter.

How can you assess your body weight category apart from weighing yourself? Choose which group best describes you?

1. Grossly Overweight

- Your body wobbles when you walk.
- You sway from side to side as you walk.
- As you walk, your legs and knees rub together.
- You have deep skin creases (wrists, arms, legs, groins, trunk and armpits).
- You have an apron of fat hanging from your lower abdomen. (You may not have seen your groins for years!)
- You have skin folds and creases that are three-finger-widths deep or more.
- You have handfuls of spare fat.

- You have no bones visible, except for those of your hands, knees and ankles.
- You have a rounded face with no visible jaw line; it merges with your neck.
- You have extra chins.
- Excess fat on the side of the face, has raised your earlobes.
- Your face and neck may merge into one.
- You may have large arms that ripple when you clap your hands.
- Breathlessness may occur when you walk upstairs.
- Your lips may turn blue after a little physical effort (an unusual finding).
- Your dainty hands and feet may be out of proportion to the rest of your body!

2. Moderately Overweight

- Your body wobbles slightly while walking (definitely on running).
- You have some skin creases present. You can 'pinch an inch' or two, somewhere on your torso.
- Your groins may not be visible without separating the creases (1-2 finger widths deep).
- Your face may be larger than before (refer to your old photos).
- You are breathless when walking at a modest pace, and after one flight of stairs (12 steps, or so).
- You are breathlessness after running 20 meters or yards.

3. Slightly Overweight

- Your weight problem may only be cosmetic, but it concerns you.
- You may have some shallow skin creases (less than one finger width deep).
- You can just about 'pinch an inch'.
- You may have some visible ribs (seen to the side of the chest when you raise your arms).
- You may experience breathlessness, but only while walking fast.
- You may have to run 50 to 100 meters, or quickly climb two flights of stairs, to get breathless.

4. Ideal Weight

- Your muscle definition shows.
- Your body does not wobble, even while running.
- You cannot 'pinch an inch' of fat anywhere on your torso.
- You have only superficial skin creases.
- Many of your ribs are visible.
- Unless you have a medical reason for unfitness (recent illness), only excessive exercise will cause you to feel breathlessness (running fast for a few minutes, or repeatedly lifting weights).

5. Underweight

- Your bones show everywhere. Your facial bones, ribs, collar bones and arm bones have skin stretched over them. Your hip bones protrude. Your legs may have something in common with those of emus and ostriches.
- On fashion-show cat-walks everywhere, this body type is common.
- You may be physically weak and need to wheel your suitcases when you travel. (No problem! Top models only need to lift their hand-bags, champagne glasses, and powder puffs).

Ready for Action?

If you are grossly or moderately overweight, do you want to take some action? Do you understand why it might benefit you to lose weight? Are you having pressure put on you by others? Are these pressures cultural, medical, or social from friends and family? Be aware of such pressures and embrace them; be prepared to defend yourself against criticism.

Have you considered why, other than vanity, you might need to take action? The best reasons are:

- To improve your health status (improved morbidity), and
- To live longer (improved mortality).

Those who are slightly overweight (Group 3 above) will sometimes see themselves as larger than they are. They should consider whether losing weight is necessary. This applies even more to those of ideal weight (Group 4), some of whom might need to lose weight to keep their job. If your work requires you to be slim, this is understandable. There are others with a body image problem, or an eating disorder, who need psychological help.

Action Alert

Those who are actually underweight (group 5), and want to lose more weight, definitely need psychological and nutritional advice. Whatever their reasons for wanting to lose weight, and to remain underweight, they are likely to have psychological issues. Many are in denial and will strongly defend their eating habits. They rarely admit to any need for help.

Eating disorders come in many forms, and not all relate to being underweight. Some are subtle; some are obvious. Some relate back to family issues and bullying. There are experts in eating disorders, and most doctors will consider referring patients to a specialist with experience.

Is Obesity a Social Problem?

This is not a subject that need interest those who want to lose weight.

Along with other western nations, the average weight of the British population is growing year by year. Some believe that the bodyweight of adults (33% of them), and children (20% of them), will soon be enough to reduce their life expectancy. Those interested in losing weight might gain some motivation from this, but I doubt it. What happens on average, and to people other than ourselves, is unlikely to provide them enough motivation needed to lose weight. Those interested in statistical analysis might ask, 'Who is to say that the risks of obesity apply to me?'

Let the government worry about national statistics; their societal role is to find the resources necessary to care for those who 'suffer', and those who are obese. Because statistics are valid only for large numbers, you and I as individuals can sideline them, except for noting the general trend and the significance that they might have for us.

Statistical calculations work only for large groups so all we can say with confidence is that a large group of overweight people has a greater risk of dying than an equally large group of people of ideal weight. So can any group data ever apply to you or me with certainty? It cannot, except by chance.

Because statistics are based on large numbers, anecdotal information about you and me, can never be extrapolated to any large group of people. Statisticians happily deal with forests; they will not comment on individual trees. Because all human beings are unique, submitting us to an averaging process discards some of our individuality.

The misuse of medical statistics abounds. They can be used to worry people unnecessarily, and unrealistically warp a true assessment of personal risk, whether it concerns bodyweight or any other medical issue.

Statistics cannot be valid (as all statisticians will tell you) when the group studied is too few in number. An individual anecdote like, 'My grandmother weighed 25 stones and lived to 95', has no statistical relevance. Anecdotal information represents insufficient data. An important point is this: if you have your grandmother's genes, you might well come to weigh 25 stones and live to 95. Who is to say you won't; certainly not a statistician or doctor.

'Surely this paradox will not have escaped my doctor's attention?', I hear you say! I'm sorry to disappoint you. What we doctors believe to be the best advice we can give for any individual, we base on large group study data. It's all we have to rely on. While we are proud of the fact that our advice is 'evidence-based' (based on statistics), doctors can easily ignore the fact that their advice applies only to large groups, and never to individual patients (except by chance). Our fallback position is this: this is the best we can do.

There is another way to practice. Before the use of 'evidence-based' information, doctors relied on their personal knowledge of each patient (valuable meta information). Many patients are lucky these days if they see the same doctor twice. So how are doctors to practice personalised medicine, without knowing their patient's personally, or having their genetic profile?

For a group of people, 35-years of age and over, obesity is not their biggest danger. Having coronary heart disease, high blood pressure and diabetes, each represent a greater risk. Contrary to the impression given by the media, we more commonly find these risky conditions amongst those of ideal weight.

At this point I must refer to an old chestnut. Some medical professionals will say that I am giving the wrong impression. Some will say that those reading and accepting what I have written, will come to ignore the serious significance of obesity. Are our patients that foolish? Perhaps these politically oriented critics were paternalistic nannies in a former life. Spin, false news, and misleading statistics are everywhere, and all those capable of thought are fully aware of it.

My colleagues will counter my views on obesity by arguing that they are not in the best interests of the majority. Some overweight people might use my point of view, NOT to lose weight. I agree with them, but I will expect them to tell their patients to beware of any advice based on group evidence.

Anecdotally (a word now used to denounce scientific validity), doctors working in emergency admissions soon come to realise that only a few presenting with heart attacks or strokes, are overweight. Politicians, however, cannot ignore the group evidence suggesting that a slimmer nation would be a healthier one (and much less costly). The science supports this conclusion.

Doctors of my generation, more than younger ones, can be uncomfortable with their patients gleaning information from the Internet. A little knowledge can be a bad thing, but things have moved on. Open public access to uncomfortable facts, may not always be politically expedient, but will help understanding and the slaying of myths. The public might discover that the status quo of society is much more important than the plight of any individual. They might find that lawyers are more concerned more with legal procedure than they are with their clients' welfare. They might find that doctors give more credence to statistical evidence than to what they can learn from their patients.

Obesity the Killer

According to many media sources, obesity is itself a killer, and not just a contributory factor. Some of it is fake news. It is heart disease and cancer (with or without obesity), that compete for the position of top killer in the western world. There are, therefore, disease processes more important to ill-health and longevity, than obesity alone. While gross (morbid) obesity, undoubtedly increases the future group risk of heart disease and cancer, the genetic antecedent risk is a key issue. For this reason, I have suggested in this book, that every weight loss program should aim to protect the heart and circulation.

I dedicated this book to a good friend of mine who was both overweight, and had coronary artery disease. He died from a sudden heart attack. See 'Acknowledgements and Dedication.'

If you have inherited heart disease (note your family history), and continue to eat foods that could aggravate the 'furring' process in your arteries, raise your blood pressure or promote the onset of diabetes (or worsen it), you could be at much greater risk of heart disease than from obesity alone. Eating protective foods could help. If you have not inherited heart disease, your weight and the food you eat, are unlikely to be of significance to your future heart health, and survival into old age.

Technical Ways to Judge Weight

This section may not be of interest to all those wanting to lose weight. Only those interested in the technical aspects of measuring bodyweight, and its relationship to illness, mortality, poverty and education, are likely to find this topic interesting.

Based on both UK and US statistics, one can refer to average bodyweight (and their ranges) that vary with age, height and ethnicity. They should respect body frame type, but do not always. Ethnicity is important because Asian, Black African, Afro-Caribbean and Middle-Eastern people have a greater chance of ill-health at lower average bodyweight.

Take a 40-year-old whose height is five feet eight inches (173cm). The 'ideal' weight for a woman might be 140 pounds (10 stones or 64 [57–70] Kgs), and 148 pounds (ten stones eight pounds or 68 [63–77] Kgs) for a man. The range of weights thought to be 'ideal' is large.

Because height makes a difference to the distribution of body fat, the calculation of a Body Mass Index (BMI) was devised. The BMI is defined as the bodyweight in kilograms [W], divided by the square of the height in meters [m]. BMI = W/m2. This is now the basis for classifying people as ideal, overweight, obese and extremely obese (see Appendix 1). Between 18.5 and 24.9 Kgs/m2, is regarded as an ideal body weight. This means that at a height of 5 feet 4 inches (1.63m), your BMI will be normal if you weighed between 7 stones 10 pounds (49Kgs), and 10 stones 6 pounds (62.2Kgs). Those over 10 stones 6 pounds, at that height, would be classified as overweight. At six feet tall (1.83m) you would have to weigh 13 stone 2 pounds (83.5Kgs) or more, to be classified as overweight (Class I obesity).

For those 5ft 8 inches tall (1.73 meters), the respective BMIs for those classified as overweight, obese and extremely obese are:

- For Class I obesity (overweight): a BMI between 25 and 30, and a weight between 75 and 90Kgs (11 stones 11 pounds to 14 stones 2.5 pounds).
- for Class II obesity (obese): a BMI between 30 and 40, and a weight up to 120 Kgs (between 14 stones 2.5 pounds and 18 stones 12.5 pounds), and,
- for Class III or extreme obesity, a BMI greater than 40, and a weight over 120 Kgs. (over 18 stones 12.5 pounds).

There are many on-line calculators for BMI. Remember to measure your height in meters and your weight in kilograms. If you want to calculate your own BMI, refer to Appendix 1, or Google 'NHS

Calculate your BMI'. The BUPA, BMI calculator is also an easy one to use.

Mortality & Obesity

The Results of Research

In one large study, the lowest risk of death (subjects followed-up for 20 years, in those who have never smoked), had a BMI between 20 and 24. (British Medical Journal: 2016; 353: i2156. doi: 10.1136/bmj.i2156).

The relative risk of death (follow-up over 20 years), increases from average (with a BMI of 25), to over three times the average, when the BMI is 40 or over (extreme obesity). As bodyweight grows further, the risk of death grow even further (in a straight line fashion).

Let's look at the risk of death for each overweight group, compared to those of ideal weight:

• For those with Class I obesity (overweight, BMI between 25 and 30), the risk increases 1.4 times (a 40% increase).

• For those with Class II obesity (obese, BMI between 30 and 40), the risk is 2.5 times greater than average, more than doubling the risk of death.

• For those with Class III, extreme obesity, and a BMI greater than 40, the risk of death is three times more than average. It becomes five times more than average for those with a BMI of 45 and over.

Being underweight, albeit not the focus of this book, also doubles the risk of death and illness.

Notice that these figures relate to 'the risk of' mortality, without further specifying the detail (the meta-data) that would link it to you and me. Because this study was extensive, we can take these figures as a reliable guide, albeit with many caveats for individuals. The results provide invaluable information for those inclined to preach to us, telling us how we should lead our lives, and what our weight should be.

From 1960 to 2003, the average US male increased his bodyweight by 15 Kgs (33 pounds); for women the increase was 11 Kgs (24.2 pounds). Increases in BMI have occurred worldwide, varying from one country to another. Over the same period, US men and women have gradually used fewer calories while exercising (140 Kcals less in men; 80 Kcals less for women).

Now for a mystery. Many long-term population analyses have related increasing body weight to an increase in death from cardiovascular disease. In fact, there has been a 60% reduction in cardiovascular deaths during the same period that saw a gain in average national body weight. (See: National Heart, Lung and Blood Institute. Morbidity & Mortality: 1998). The link between the two is broken, but why?

Over the same period, there has been a lower consumption of saturated fat (fatty meat for instance), an increased consumption of fish, fruit and vegetables (a Mediterranean diet), a greater use of vitamins and minerals, and fewer who smoke. Could these factors be of more relevance to heart attacks than obesity?

Simultaneous changes in education and poverty will also influence the figures. There are interesting relationships to be found worldwide between poverty, education and bodyweight (BMI).

As showed by the US National Health and Nutrition Examination Surveys (2011), the more years spent in education (>16 years compared to < 12 years), the lower is the average BMI (14 points lower for women; 11 points lower for men). As we pass from the lowest income group, to one four times greater, the BMI drops (10 BMI units lower). The poor and less educated, have higher BMIs than their richer and better educated fellow US citizens.

For more detail see Bibliography: Shook et al. 'What is Causing the Worldwide Rise in Body Weight'.

CHAPTER 6

LOSING WEIGHT. HOW SIMPLE IS IT?

I believe politician Anne Widdicome once refused a request from her publisher to write a book on weight loss. Perhaps she thought the task too simple for a book. She recommended that fat people should spend less money on expensive dieting fads, and simply eat less and move more. (Ann Widdecombe Versus the Diet Industry, ITV, September 2008).

I started writing this book once I realised that her scientifically valid, but abbreviated advice, would not help those who repeatedly tried to lose weight, and had failed repeatedly.

Advice is easy to give. 'Get fit'; 'Love thy neighbour as thyself'; 'Take it easy'; 'Work hard and play hard'; 'Forget it /him/her, and move on'; 'Eat less and exercise more'; these are all bits of advice, easier said than done. Wise advice can add perspective to a problem and encourage people to make beneficial changes.

Is eating less, and exercising more, all one needs to know about losing weight? It is for many. They are those who need no further advice, and do not need this book. That leaves a large minority who cannot easily say 'YES' to the following questions:

Q1: Faced with a pack of six doughnuts, filled with delicious looking raspberry jam, are you able to eat just one, and leave the rest?

Q2: Do you eat only when you are hungry?

Q3: Have you failed to lose weight by eating less, and moving more?

Here are some questions I have been asked by patients who failed to lose weight:

- How can I eat less and not get too hungry?
- Which foods should I eat? (are some foods better for me than others).
- How much food should I eat?
- Which type of exercise is best for me, given my weight and fitness level (which type, how much, and how often)?

The answers to these questions, and the exact details of 'how to', are what many need. It would be easier of course for them not to bother with diet and exercise. Surely there's a pill, injection or operation that would work without the need for a struggle. Surely there must be an alternative to the 'no pain, no gain' principle - the Holy

Grail of weight loss. Some pharmaceutical companies know this, and realise that weight loss made easy, would be big business.

With genetic analysis, and research into the gut microbiome and gut hormones, more is now known. Discovery of the Holy Grail of weight loss could be getting closer. You will find more detail in Chapter 19.

In the minds of many overweight people who have repeatedly failed to lose weight, what went wrong can elude them. The reason for their repeated failure can seem mysterious.

In looking for the causes of being overweight, few will look into the fuzzy realms of mind-set, culture, and attitude to food, eating and exercise. Many are aware of the role played by self-gratification, self-discipline, and the adverse effects of stress, but they find it easier to limit themselves to counting calories.

There is one group of people who regularly annoy all those seeking weight loss. They don't need to eat less and move more. They seem to eat as much as they like, do almost no exercise, and never gain weight. Why might this be? They possibly eat very little and exercise more than they say. Perhaps they worry a lot more than most, and burn a lot of calories doing it. Might cancer or coeliac disease explain it? Perhaps it's their metabolism.

While I was studying for my 'A' Levels, a student colleague of mine was Mike Stock. He was later to become Professor Michael Stock, a nutritional science expert. He discovered brown fat; fat which burns much more energy than the ubiquitous white fat.

He put athletes and unfit people into a sealed room. He measured every calorie that went in and all that came out. The unfit people put on more weight than the athletes, despite eating the same number of calories. Using thermal detection cameras, he found that athletes radiated more heat from between their shoulder blades than the unfit subjects. Anatomists then found brown fat in this location. The athletes didn't put on weight because, even sitting on a chair, they lost more calories than their unfit fellows.

Is Being Overweight Useful?

Darwin's work revealed how the survival of the fittest arose from advantageous evolutionary adaptations. It prompts the question: has there ever been a biological advantage to obesity?

In evolutionary terms, prehistoric humans with an excess of stored fat, might have better survived famine. Obesity might have favoured the endurance needed for slow, long distance walking, (to where the food was) when local food was in short supply. Thin prehistoric humans on long-distance journeys would need to feed regularly. In the western world, this advantage of obesity no longer exists, although famine still occurs regularly in some other parts of the world.

Inducements to Eat

No law yet bans obesity. Society is content with shaming and blaming the obese. No law yet restricts either what we eat, or how much we eat.

Some cultures foster eating. Obesity is alive and well in all western societies, sustained by cheap carbohydrate-rich food, available everywhere. Money-saving offers, now encourage us to eat more. In some restaurants one can 'Go Large', or get 'King-Size' portions; both provide better value for money (cheaper per calorie).

Advertising can induce us to buy certain food products. Food manufacturers know that many find it difficult to resist a bargain. Subtle advertising, specially reduced prices, and effective store lighting, can highlight the foods they want to sell. The many inducements on offer, can easily influence choice. There is now more to choosing food than simply keeping to our usual preferences.

Fitness is Still Fashionable

With health professionals energetically advising fruit, vegetables and exercise, those who are fit and slim are liable to become sanctimonious. They are 'in fashion' and admired for their slim appearance, firm six-packs, and physical prowess. That they are more attractive and desirable to others is nothing new. In ancient Greece, the Spartans extolled the same virtues.

Binary thinking has it that every 'Ying' must have a 'Yang', but there are many historic examples of 'Ying' at war with 'Yang'. The fitter Spartans despised the decadence and hedonistic lifestyle of the ancient Athenians. Perhaps they could have been a bit more sympathetic. After all, our mentality and lifestyle arise from our intrinsic nature and cultural influences, both of which are largely beyond our control.

Because our social and family group culture has nurtured us, attempts to change our outlook and mentality are liable to be viewed negatively. The working-class boy who tries to become a ballet dancer, can easily lose his place among his friends. Our attitude and lifestyle are not always matters of choice; the cultural influences upon us are coercive and replete with demands and expectations not of our choosing.

Many claim to be fit and happy; others claim to be happy with their obesity. I have met many smokers and obese patients who readily accept the idea that dying a few years earlier than predicted, would be a reasonable price to pay for the pleasure their chosen habits give them. For the fit, dying a year or two later than predicted might not seem a sufficient reward for all that exercise and food restriction. In addition, the physical toll that exercise exacts on joints, and our other tissues, needs to be accounted for.

While the energy, health and extra agility gained from being athletically fit is accepted as being worthwhile, the obese need our sympathy: they may never have experienced fitness, and may never have appreciated its merits.

The Resolve to Lose Weight

When trying to lose weight myself, the most difficult question I had was:

'Where do I find the resolve to stop my excessive eating?'

How was I to

- Stop being seduced by food?
- Stop eating when not hungry?
- Stop eating when I had nothing better to do?
- Stop eating as if I feared famine?

After successfully losing weight over six weeks by simply reducing my calorie intake, I visited the British Library to read what others had written about the subject. I read not only the popular books, written to sell a dieting regime, but also the results of scientific experiments undertaken to define the best weight loss methods. It soon became obvious that there were reliable weight loss methods that were not common knowledge.

According to Occam, if the answer to a question isn't simple, it's likely to be wrong. It was gratifying to find a few simple answers to my weight loss questions.

If you have failed to lose weight, despite trying many dieting regimes, consider this question: 'Have those whose aim is to make money from selling you diet plans, dieting books, and food counselling, led you up the garden path?'

Losing weight requires less time and energy than the endless search for a magic diet, falsely promising weight loss without a struggle.

Here are some factors to consider when trying to lose weight, and trying to maintain the weight loss successfully. My aide memoir is:

PIE FACES

P is for Portion Sizes.

I is for Image – your Body Image.

E is for Eating. Why do you eat? What do you eat? What should you eat for a healthy heart?

F is for a 'Fat' or 'Fit Mentality'.

A is for Appetite Control (Genes. Gut hormones and brain chemicals).

C is for Carbohydrates, and Control over what foods you eat and portion sizes.

E is for Exercise.

S is for Smoking.

Medical Ethical Alert

We all need to decide how we want to lead our lives, what we want to eat, and what weight we prefer to be. At one extreme, there are those who love their food. They may be overweight, but are content with the way they feel (they may be at risk, but see no need to acknowledge it). At the other extreme, are those who are underweight, abstemious with food and drink and proud of their discipline. Both are more likely than those of average weight to die earlier than expected. 'Likely' should never imply 'necessarily' to any individual.

A Doctor's Viewpoint

For both overweight and underweight people, counselling and nutritional guidance can be useful. We should encourage smoking cessation, because it has many long-term health benefits. Unfortunately, giving it up is often associated with weight gain (nicotinc increases metabolism, decreases appetite, and helps dissolve fatty tissue [lipolysis]). Smokers know this and continue to smoke to minimise weight gain. Assuming that nobody wants to suffer illness, and a few may actually want to die early, both smokers and the obese need to answer the same question:

> 'How long do I want to live?

If doctors knew the longevity expectations of their patients, they could better advise them about following effective health promoting strategies. Whether their patients will find these strategies acceptable is another matter.

When giving medical advice, the words 'should' and 'need to', are best avoided. To employ them without some knowledge of each patient's desire for longevity, may be well-meaning, but paternalistic. Physicians practicing the art of medicine need to transcend the limits of scientific understanding, and gain much more personal information about their patients. To practice the art of medicine effectively, doctors need to understand their patients as intimately as they would a close friend or family member. This suggestion will undoubtedly make some doctors cringe; doctors identify as scientists, not care-workers or counsellors. A doctor's professional work description is consistent with them delegating patient care to nurses, social workers and family members. Although doctors must give priority to the objective evaluation of disease and the treatment of patients, it is a pity that those who manage clinical organisations, have chosen to reduce the time available for pastoral care and the art of medical practice.

If they are to have acceptable patient doctor experiences, overweight patients need to be aware of the type of doctor they consult, and their orientation to weight problems.

A Patient's Outlook

Do you agree with these statements?

• Happiness and contentment are more important than body weight.

• Happiness and contentment are more important than length of life.

• Based on large-scale group studies, excessive weight gain is best avoided by those who want to preserve the length and quality of their life.

• Health, fitness and an ideal weight, favour improved average longevity (for a large group of people), but no guarantee of longevity can be given to any individual within a group.

• Heart attacks and high blood pressure occur frequently in those of average weight; those who are non-diabetic, and those who have a normal blood cholesterol. For these reasons doctors are liable to over-rate obesity as a factor when predicting disease and the death of any individual patient.

Not every risk factor is all bad. Smoking, for instance, was once used to cure asthma. It certainly aids weight loss and improves mental concentration. Unfortunately, it can trigger cancer and coronary artery disease, with only one third of smokers living to 65-years of age.

Many are comfortable with obesity as a lifestyle choice, even knowing that it is a risk factor for diabetes and heart disease. They may come to accept this as a price worth paying.

Chapter 7

I'm Overweight. Whose Fault Is It?

In 2011, I attended a lecture at the Royal College of Physicians, London. Giving the Croonian Lecture, a doyen of body weight physiology, Prof. Stephen O'Rahilly (Cambridge University), revealed that there were still many mysteries about weight control to solve. Since then, some fascinating facts have emerged.

Those who have tried to lose weight and have failed repeatedly, might ask, 'Am I overweight because of a fault in my genes?' or, 'Is it my metabolism the problem?' Science has advanced, and inheritance, gut hormones, and brain chemicals, are now known to play an important part in appetite control, the feeling of fullness, and weight control. Surprisingly, some genes can direct the food choices we make.

Genes & Jean Size

Have you ever been swimming, and noticed Dad and daughter, or Mother and son, getting into the water? Has it ever struck you they have a similar shape? Could this observation be evidence of an inherited trait? The effect of some genes may only be small, but some predict body shape and weight.

Consider addiction. Drug addicts more often possess the genetic (transcription) factor DeltaFOSB, than those without an addictive tendency. Could this genetic factor affect our approach to food, and the interest we have in doing exercise? Both now seem likely.

The MC4R gene has variants, some of which can increase our desire for high calorie food, while others affect appetite and overeating. The effect may be small, but a small change in appetite over many years could eventually bring about a big change in bodyweight.

The desire for high calorie, sugary and fatty food, has been linked to the FTO gene. Genes and the brain chemicals they influence (dopamine and serotonin), are involved in the reward we feel when we eat certain types of food. Both dopamine and serotonin can impart the feeling of pleasure. Is this why some people feel happy when they eat certain foods? Some anti-depressants work in a similar way; serotonin can be made to persist for longer in the brain. Some have implicated dopamine in food lust and an inability to say 'No' to food.

One interesting way to gain an insight into genetic influences is to study twins; especially those who were separated at birth and came to lead different lives. One study of twins who were reared apart, showed that their body weight remained the same as they grew older (a 70% correlation). This implies that their genetic makeup had more to do with their body weight control, than their lifestyle and cultural influences. Our inherited genetic influences can affect, not only our food preferences, but also how we choose to spend our energy. (See in Bibliography: Stunkard et al. 1990, for more information).

The update of The Human Obesity Gene Map report of 2005, identified one hundred and twenty-seven obesity associated genes, twenty-two of which were found influential in five separate scientific studies. The conclusion was this: genes count when trying to explain obesity. Exactly how they do it, remains to be discovered.

What we know is that genes help to control appetite, hunger and the feeling of fullness. They can influence gut hormones, brain chemicals and our individual psychology. Their influence might be small, but over time that can make a big difference.

Genes and Mentality

Do genes play a part in the development of a fat or fit mentality? This is likely, given that certain inherited behavioural characteristics, are typical of different species. Koala bears like sitting in trees eating eucalyptus leaves; polar bears roam around on the ice, trying to catch seals.

Because the genes mentioned so far, influence both our behaviour (what we choose to eat) and our metabolism, they might also be associated with the genes that determine our mentality. If this is so, we should expect to find variations, like obese people with a 'Fit Mentality', and those with an ideal weight who have a 'Fat Mentality'. This is in keeping with known biology: many variations allow the selection of those fit for survival.

We should expect biological variation to allow for those of ideal weight, who have a 'Fat Mentality', and those with a 'Fit Mentality', who eat sparingly, and remain overweight.

The Butterfly Effect

Our genes do not exert a strong influence, but small genetic differences can result in large effects in the long term.

In the 1960s, the meteorologist Edward Lorenz was studying weather prediction using computer models. He discovered that small changes in the initial specifications, could cause major differences to the predicted weather. When he made small changes to the initial wind speed or air temperature, the model predicted a completely different weather pattern. These small differences could change the prediction from peaceful weather, to violent storms. He named the influence of these small changes (like a butterfly flapping its wings) 'the butterfly effect'.

The butterfly effect can work to increase or decrease the outcome. Reducing or increasing our calorie intake by a small amount each day, could have a large effect on our body weight over a one year period.

My friend Frank loved sugar, and wanted to lose weight without following a strict dieting regime. The only action he took was to cut down the number of spoons of sugar he took in his cups of tea. He reduced the amount from three teaspoons, to one teaspoon per cup. From this intervention alone, he lost 20 pounds in the first year. The key to his success was his long-term commitment. Unlike many people, he had no need for rapid weight loss.

New Hormones and Chemicals

A detailed study of this subject is not possible here. Human metabolism is too complicated. It is possible, however, to provide a simplistic overview of the subject.

Leptin is a substance produced by fatty tissue. When levels in the blood are high, the brain switches off its starvation mode (which would make us eat as much as we can). The result is to signal the need to eat less, and to spend more energy. It can, however, spark off a complicated cascade of chemicals within the brain (neurochemicals). These sometimes increase appetite, although when acting specifically in the hypothalamus, they can decrease appetite.

Genes control the release of substances like ghrelin (a hormone produced by the stomach), GLP-1 and YY peptide (all substances that are made in our lower small intestine). They are all involved in feelings of satiation or fullness.

Ghrelin has been called the hunger hormone because it can stimulate appetite. It can influence other hormones like insulin and growth hormone. Peptide YY has been called the satiety hormone because it reduces appetite and promotes the feeling of fullness.

When GLP-1 remains low despite eating, hunger can persist. Filling of the lower intestine stimulates its production. A high fibre diet takes longer to digest, and can reduce hunger by stimulating GLP-1 production. This is because fibre lasts long enough to fill the lower small intestine. A feeling of fullness results, even if eating less food.

By understanding more about genetic and hormone factors, it should be possible to design specific weight loss strategies for each individual. That is for the future. For now, the short take-home lesson is that eating fibre can help to control our appetite.

Traditional Hormone Problems

A few overweight people have an underactive thyroid; a small number will have an overactive adrenal or dysfunctional pituitary gland. Female hormone production changes during the menopause, and synthetic hormones can be used to replace them. The contraceptive pill adds extra hormones to the normal hormones of younger women. There are the implications for women trying to lose weight.

As well as promoting body fat accumulation, an excess of female hormones can cause fluid retention. During the menopause, some women take replacement female hormones to prevent hot flushes. Many notice an associated weight gain. When we measure female hormones in the blood, the results do not always explain the symptoms. Individual sensitivity to female hormones is one factor. At worst, the menopause can be difficult to cope with. At best, it can pass without notice. Losing weight during the menopause, or while taking the contraceptive pill, can be a challenge for women.

Medical conditions are infrequent causes of obesity, but onsider an underactive thyroid. This has several specific characteristics. Thyroid hormones help to control our temperature and influence how hot or cold we feel. There are two thyroid hormones, referred to as T4 and T3 (the number refers to how many iodine atoms present). When either is deficient (usually T4), the most prominent symptom is to feel cold when others are warm. Some patients will want to wrap up in warm clothes, even in summer. Other less specific symptoms are weight gain, hair loss, and a dry puffy skin. A doctor can confirm the diagnosis by examining you, and measuring the levels of thyroid (and pituitary) hormones in your blood.

An endocrinologist colleague of mine at Charing Cross Hospital, once asked me to undertake some research. He wondered whether coronary heart disease ('furring' of arteries) might be related to thyroid metabolism. This did not prove to be the case.

When I tested the thyroid function of some of my overweight patients (using thyroid stimulation), the suggestion that their obesity had a hormonal cause, gave some the hope of an easy cure. I found only two cases of a sluggish thyroid. Both lost weight easily after taking a small dose of thyroid hormone. Taking extra thyroid hormone when the thyroid is not under-active (or sluggish on stimulation), is not an advisable treatment for weight loss. Thyroid hormones have too many side effects.

An adrenal gland tumour, producing extra cortisol hormone, is a very rare cause of obesity (Cushing's Syndrome). Doctors do, however, prescribe cortisone for many medical conditions (for asthma and immune conditions). This can cause weight gain. In over 50 years of practice, I never saw one case of Cushing's Syndrome; endocrinologists manage most of them. Extra cortisol in the body causes easily recognised bodily changes: the face becomes rounded (moon face); the torso and abdomen enlarge, as the legs become thin from muscle wasting.

Pituitary gland problems are rare and are only rarely responsible for weight gain. Pituitary gland hormones control most of the others. An increase in each pituitary hormone causes a specific syndrome. An overactive thyroid (occurring from thyroid disease or an increase in pituitary thyroid-stimulating hormone) can cause weight loss, palpitation, nervousness and tremor.

Extra male hormone (testosterone), and extra female sex hormones (oestrogen and progesterone), can cause weight gain. This is sometimes a feature of polycystic ovaries, the other features of which are lower abdominal pain, precocious (early) puberty, male pattern hair growth and menstrual disorders.

Extra growth hormone in adults will thicken the skin and broaden the nose, resulting in the typical facial appearance of acromegaly. In children, extra growth hormone will cause a rapid gain in height.

Mindset

Have you heard a little voice in your head saying,

> 'OK. You shouldn't have eaten that!
> Your self-discipline failed this time.
> So what? Why worry?
> Eat and enjoy it.
> Life is too short to worry about your weight!'

By now you should be aware of the stranglehold your mentality can have over your food choices, the amount you eat, and the exercise you do.

A 'Fat Mentality' is not my problem

Climbing Mount Everest and losing weight, have something in common: they are both difficult to achieve. If you have tried either, you will know what I mean.

Did I hear someone say, 'You cannot compare climbing Everest, sailing round the world single-handed, and walking solo to the South Pole with losing weight.' I agree, except that losing weight can be more difficult psychologically.

There is another consideration. Those with enough self-esteem, may have no need for the rewards of physical endeavour and team effort. Losing weight is intensely personal, and collaboration with others does not always help. Losing weight is a personal battle; a fight between our fit and fat mentality. Whatever the endeavour, failure can make victims of our vanity and self-esteem.

Losing weight for some, is equivalent to climbing a very tall beanstalk, the aim of which is to defeat an ogre and run off with a goose. Like Jack, you must be fit enough to climb a beanstalk, and be capable of defeating a giant ogre. You might then capture a goose that will lay golden eggs for you.

The metaphorical ogre in this case is our 'Fat Mentality'. It will do all it can to dissuade you from climbing anything; there is too much effort involved. Those who climb Everest, sail around the world single-handed, or walk solo to the South Pole, get to be newsworthy; some win medals and get knighthoods. If you climb a beanstalk, beat your own personal ogre, and get some golden eggs (lose weight successfully), few will get to know about it. You will, however, have earned a reward - feeling good about yourself.

Achievements are all relative and depend a lot on cultural outlook. Walking to the South Pole is nothing compared to a paraplegic taking her first step. Losing weight for those with a 'Fat Mentality', is ten times more difficult than it is for those with a 'Fit Mentality.' If you are overweight and happy about it, smile and chuckle benevolently whenever you hear the cop-out: 'losing weight is easy. Just eat less and exercise more!'

Not everyone has a weighty ogre to defeat. They might just have been a little greedy over Christmas. Those with a 'Fit Mentality' in charge, can easily lose weight by eating less and exercising more. It's easy for them.

I don't want those with a 'Fat Mentality' to be beaten by an ogre, whose only desire is to make you fail when trying to lose weight. My advice is to go no further until you have learned how to defeat your 'Fat Mentality'. Read on to find out how.

PART THREE

THE DIETING PYRAMID

CHAPTER 8

CLIMBING THE DIETING PYRAMID

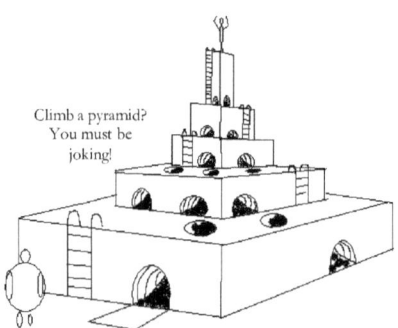

Figure 1: The Dieting Pyramid

After understanding the mentality that drives our eating and exercise, a few further practical steps are essential for successful weight loss. Many experienced slimmers think they know them, but have still failed to lose weight. Like climbing the step Pyramid of the Sun at Teotihuacán, Mexico, a progression of steps faces those who wish to reach the top. Each step is simple to describe, but challenging.

Each step in the process of weight loss involves gaining some knowledge. A change of attitude towards food, eating and exercise, may also be necessary to better control eating and exercise behaviour.

Is The Time Right for You?

After losing his job, a 55-year-old patient of mine, decided it was time to lose weight, give-up smoking, and take up squash.

His decision raised a few eyebrows and prompted a few questions. Why would a stressed, unhappy person, want to give up smoking and curtail his eating habits when he was facing so many problems? Could it be an egotistical need to prove his true grit?

Perhaps he hoped to prove that he could succeed against the odds. Perhaps his survival instinct had kicked in, and he needed to get fit enough to deal with the troubles ahead.

He could have taken another route. Having felt stressed and depressed, he could have given up all endeavour. After repeated personal failures, some assume that they are certified failures, and cannot face trying again.

One grossly overweight patient of mine once told me about her repeated failure to lose weight. 'I'm not depressed', she said, 'I've not lost my discipline, and I just don't care anymore!'

There are many with failing discipline, who have no wish to fail. They put off challenges, including weight loss, until . . . mañana . . . bukhara . . . demain . . . zaftra . . . or domani. Whatever you call tomorrow, a degree of self-deception is involved. 'I'll do it when I'm ready', they will say. What they actually mean is, 'I'll carry on with what I want to do for now, and pacify my critics with a plan of action'.

This approach can be wise under certain circumstances. When fully stretched by coping with many problems, why stress yourself further with dieting? We are most likely to succeed with any task if we accept how much time and energy is really required to complete it. Never set yourself tasks that are beyond the time and energy you have available. Wait until you have the required time and energy available, or you may set yourself up for failure. It is best to know your true capabilities, and avoid failure by not trying to overreach them.

If you think that you have the resolve, and the time is right for you, you should be ready to proceed and find out what is required for successful weight loss.

The Law of Diminishing Returns

This will be bad news for some. The fatter we get, the slower our movements become, and the less weight we will lose.

Ask your children to run upstairs and retrieve your mobile phone. How fast do they move when incentivised? Could you move like that? Movement burns calories; minimising movement saves energy. Unfortunately, the slower we move, the more quickly we gain weight. Cars burn most fuel while accelerating. That is why the start, stop, racing around behaviour of children, helps them avoid obesity. The more they sit in front of a computer, the more weight can gain. Physical factors alone can explain why some change weight.

There is some good news. That is the law of increasing returns. The quicker you lose weight, the quicker you will move, and the quicker you will lose weight. The more weight you lose, the more active you can become. This is something to look forward to, but only for those prepared to switch from a 'Fat Mentality' to a 'Fit Mentality'.

Take another look at my weight loss pyramid. The going will be slow to start with, given that every climber is overweight, slow to move, and poorly motivated at this point. More activity will often be undertaken after some weight is lost. As confidence grows, progress becomes faster.

The Task Ahead

- To get to the top of any step pyramid you have no option but to take one step at a time.

- There are no short-cuts, although, there are ways to make the going easier.

- Like climbing Mount Everest, you will need to stop many times to refresh yourself, and to acclimatise.

- Unlike climbing Everest, there is only one way up the weight loss pyramid.

- Impatience will set you back.

- Jumping ahead is not an option, however frustrated you feel.

I guess this all sounds tough. You're right. It is! Unfortunately, that's the way it is. Many people find it difficult to lose weight because of the difficulties they know they must face. Better to be fully aware of the difficulties and how to overcome them, than believe false weight loss promises.

Two Vital Questions

Before climbing Everest or attempting to lose weight, ask yourself these vital questions:

1. Is this what I really want?

2. Is anyone pushing me to do this?

- If you are doing the pushing, you stand a chance of success.
- If someone else is doing the pushing, reconsider what it is you want.
- If it is someone else pushing you, ask what's in it for them, and what's in it for you.
- Don't tolerate emotional blackmail like, 'I can't love a fat person!' or 'I'll leave you unless you lose weight!'.
- Don't accept without question arguments like, 'You'll feel better after losing weight.' 'You'll live longer if you weigh less.' 'Your blood pressure won't come down unless you lose weight.' These are all cop-outs; unfair forms of persuasion, even if true.

If you agree to having your life controlled by others, that's fine. Many feel safe in the hands of a stronger person; someone they respect. As part of a loving and supportive relationship, this is understandable; as part of a manipulative relationship it is not.

Beware of Medical Arm-Twisting

Here's an arm twist, sometimes used by doctors on their overweight patients:

If you lose weight, you will lower your blood cholesterol, avoid a heart attack and lower your blood pressure (and thus avoid a stroke)!

Even though this is true ('on average' for large groups), you might ask why so many thin people have heart attacks and strokes.

Doctors can only address the risk factors known to apply to large populations. Actually, this is the best we can do, because we can never completely validate or guarantee the advice we give to an individual.

Arm-twisting has its place, but statistically based (so called evidence-based) arm-twisting, can be deceitful. More to the point, arm-twisting overweight and obese people, rarely motivates them to lose weight.

Vain Hopes

Vanity is a human characteristic. It varies from suppressed to much flaunted. When it motivates cosmetic surgery, it can lead to either despair or elation; when it motivates weight loss and improved fitness, it can lead to improved self-esteem.

Some methods used to pander to the vain are false and unscrupulous. The Emperor did not have beautiful new clothes, as his followers proclaimed; he was naked. Are there really skin creams that will make you look ten years younger? Beware of anything aimed at satisfying your vanity.

Vanity and hope can make sweet partners, but both can be followed by a bitter aftertaste. That expensive clothing item you bought to enhance your appearance, may not seem so wonderful a few days later. As we age, the vainglorious hopes some have for extending their youthful looks can intensify. The methods some will use to achieve it can be extreme. Hope killed no one, but if you are not to be saddened by the failure of medical science (so far at least) to keep you young and fit, be careful not to allow vanity to rule your life completely. Every week, one aging supermodel or the other, turns to alcohol and drugs. Could it be, that despite all their efforts, they still dislike what they see in their mirror? Those who are realistic about aging can be gracious (our late Queen and Queen Mother, Sophia Loren and Joanna Lumley, provide examples of graceful aging). Older people especially, should give thought to what they expect to achieve from losing weight.

Satisfying vanity can be expensive. If you are young and seeking a mate (like the film character Bridget Jones), improving your appearance in the eyes of others can pay dividends. Because so many potential partners are fickle, with undeclared emotional agendas, you might never win them over even if you were twice as beautiful, and half your weight!

If your key doesn't fit the lock, only a destructive force will open the door. No effort is required to open a door with a key made for it. Never entertain the idea that the attractive potential partner you fancy, will lie at your feet and beg for your attention, if only you lost some weight. In plain English, that's crap! Unfortunately, like all serious lessons in life, they are best learned from experience. Some who cannot adjust to reality, are driven to jump off bridges or to commit suicide with alcohol or drugs. Many remain depressed, while others remain composed and carry on.

If you are content with yourself the way you are, you have achieved something special. You have reached an enviable state of grace. If you have achieved this, not by pursuing selfishness and vanity, but by being kind and helpful to others, you will have achieved some success in life.

Weight loss as an aim, must assume its proper place. We can expect it to make us feel healthier; make us feel less breathless, and boost our self-esteem. It will surely enhance our self-esteem, if we have held our discipline in place, long enough to achieve a weight loss goal.

Pyramid Climbing for Beginners

If you are ill-informed, climbing any large pyramid will bring you face to face with challenges. These difficulties could depress or dishearten you. They will certainly test your resolve. You will need to rest sometimes and learn more about climbing before moving on.

Well-meaning friends may offer advice and encouragement; others may sound words of caution and question your wisdom. 'What are you doing to yourself?', they may ask.

If the goal is to lose weight, you will need a clear idea of what you want to look like. If you don't succeed, your negative friends will comfort you with, 'What you wanted was not achievable', or 'If you don't succeed, try again'. Knowing they were right will boost their ego (their psychological goal), but possibly depress yours.

Be prepared to get stuck while climbing a pyramid. Each time this happens, ask what extra knowledge or know-how, might help you up.

Look at my pyramid. Because it has wall ladders, the larger you are, the more difficult it will be to climb. After climbing each new step, you must negotiate new pitfalls; slides that can take you back to the previous level, or even back to the start of the process. For most of us, progress in life follows the same rules as the Snakes and Ladders game. Luck, and the rules of the game, are both involved.

If you have been lucky enough to stand before the Great Pyramid at Giza, you will know how awesome it is. (I hate that word, but just occasionally it is justified). Climbing the weight loss pyramid (also man-made) is no less daunting, but with the correct information, you should be able to conquer it.

A Seven-Day Plan

Here are some important points:

- Don't set yourself an unrealistic target for weight loss.
- Plan your eating one week at a time, not day by day.
- Do not weigh yourself more than once a week.

During each one-week period, you can vary what your eat day by day. Follow the principle of cutting down the carbs and calories differently each day, but more about that later.

Any failure to keep to the plan you choose for yourself, will nourish your 'Fat Mentality'.

Let's look at where you are heading.

STEP 1: The First Step.

Successfully climbing the first step means you fully understand your 'Fat' and 'Fit Mentality', and are ready to combat every attempt your 'Fat Mentality' makes to control you.

STEP 2: Appetite Control

- You must learn how best to control your appetite.

STEP 3: Reducing the Carbohydrates

- You must learn about (glycaemic) carbohydrates, and which

foods contain them.

- You must learn how much carbohydrate to eat.

- You must learn about protein and fat, and in which foods they are to be found.

STEP 4: Controlling Portion Sizes

- You must learn how much of each food you like, contains 100 Calories.

STEP 5: Healthy foods: those that might prevent disease.

- You must learn which foods are best for your general health, your circulation, and your heart.

Chapter 9

The Second Step. Appetite Control.

WHO LOSES WINS. WINNING WEIGHT LOSS BATTLES. 117

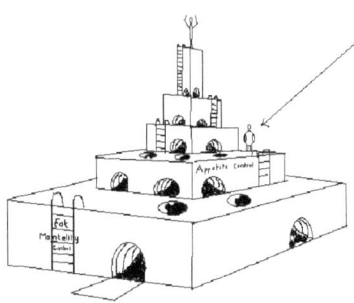

Figure 2: The Dieting Pyramid. Appetite Control

> There is no love sincerer than the love of food.
> *Man and Superman*, George Bernard Shaw, 1903.

On the second step you must eat less, eat more fibre and be comfortable with both. But how?

Before trying to understand appetite control further, here are some simple strategies that could work to curb your appetite:

1. Immediate strategy: drink lots of water. This can quickly ward off the need to eat. It also helps to prevent constipation.

2. Longer term strategy: eat less and your desire to eat will lessen.

3. Occasional strategy: use a herbal appetite suppressant, like Hoodia or green tea extract. For other options, see Chapter 19.

Do any of the following statements reflect your views?

- 'It's a sin to waste food.'

- 'I was taught to eat all I was given.'

- 'Waste not, want not.'

- 'I might as well eat everything on my plate: I've started, so I'll finish!'

- 'You never know, I might get hungry later.'

- 'No-one can survive long without food.'

- 'I can't go to work on an empty stomach.'

These statements have one thing in common: they are excuses for unnecessary eating. Nobody who is unhappy with their excess weight can afford them.

Cities in the western world now provide 24-hour cafeterias and food outlets. They make big profits servicing our self-indulgence. Coffee shops, food shops, cafes, restaurants and cafeterias line our streets, some offering food, food, food, twenty-four hours a day, every day.

If the directors of food outlets want to engorge their bank accounts, we must gorge ourselves with their food. Food companies will then have the cash to saturate us with even more food advertising. There is now so much food advertising, it is difficult to escape. Food promoters will target those who are slaves to their appetite, but should we hold them responsible for the growing prevalence of obesity? In their defence, they can argue that, 'Nobody makes anyone buy food!' Also, 'Nobody forces food into our mouth!' and, 'Everyone has a choice.'

Our appetite has to resist those whose commercial will is for us to eat as much as we desire, whenever and wherever, we desire it.

Low calorie slimming diets that induce hunger, are unpopular and mostly unsuccessful. Preferred diets are those that allow us to eat what we like. Food not only feeds our mouths, it feeds our 'Fat Mentality'.

A remarkable feature of Dr. Atkins' Diet is that he did not restrict fatty foods. Having plenty of fat to eat might feed a 'Fat Mentality', and long-term, could even hasten heart disease in those pre-disposed to it. Here, I have unfairly used the possibility of getting heart disease, as an arm-twister. It's what some doctors will do to encourage their patients to take action. In fact, losing weight with the Atkins' diet for a couple of months is likely to be beneficial, rather than dangerous, because of the associated weight reduction. Saturated fat can hasten artery 'furring' and coronary heart disease, but that is a long-term effect that takes years.

A word of warning about drawing connections between two sets of independent observations, like weight loss and heart attacks. It could be journalism, not science. Only a few journalists have a scientific training, and the scientific 'facts' some serve up and the connections they suggest, may not always be reliable.

Most overweight adults trying to lose weight, will get stuck at one time or the other. They can reach a point where they eat hardly anything, and lose no further weight.

There are two common reasons for this:

1. The amount of food I just referred to as: 'hardly anything', is likely to be more than any slim person would eat. If you never allow yourself to get hungry, this is likely to be the case.

2. The second reason is that 'hardly anything' can comprise all the wrong foods; foods that contain too much energy to burn off easily. Just occasionally, of course, 'hardly anything' might actually mean just that. We shall see.

Many overweight people say:

- "If only I didn't have such a big appetite, losing weight would be easy."

- "It's so difficult to eat less."

- "I'm used to eating well."

- "The problem is, I love bread and have a very sweet tooth."

Why not get some appetite suppressants? What's the problem? Unfortunately, those who feel overweight only rarely lose weight permanently by taking appetite suppressant pills. There is a simple reason for this. Pills will do nothing to reduce the influence of a 'Fat Mentality', and do nothing to promote a 'Fit Mentality''. Without beating our 'Fat Mentality' into submission, long-term weight loss will usually remain a dream.

There will always be exceptions. There are some overweight people (I don't know how many) who genuinely eat very little. They have a 'Fit Mentality', and follow all the advice about exercise, portion sizes, carbohydrates and calories, and yet do not lose weight. They may have a genetic, medical, or metabolism problem, that could benefit from medical appraisal and intervention.

The feeling of hunger is normal and natural, not something to be avoided at all costs. Almost the only difference between those who are overweight, and those who are not, is that overweight people eat before hunger arises. Fit, slim people, more often eat only when hungry.

Read all about appetite suppressants in Chapter 19.

Did Anyone Mention Hunger?

'I felt hungry the other morning. I wanted a cup of coffee and a biscuit. Somehow, one biscuit wasn't enough, so I ate five; not because I was hungry, but because they were there!'

Tip: keep your biscuits and chocolate out of sight.

Overweight people only rarely experience hunger; they eat mostly before the feeling arises. Is it a subconscious fear of hunger, or an obsessional eating habit that drives them to eat?

Freud thought that what motivates us most is the pursuit of pleasure. He called this the pleasure principle. The desire to avoid pain, discomfort and hunger is similar. Is hunger (or the contemplation of it) so uncomfortable that it drives some to eat before they experience it?

Did your mother ever say, 'I've made you some sandwiches. I don't want you to go hungry.' After all, what sort of mother (with sufficient resources at hand) would let her children go hungry? Parents who want to show that they care, can easily transfer a fear of hunger to their children. There are, of course, many other ways to show affection to a child other than feeding them.

Having lived through two world wars, my father and grandparents must have known hunger. It was not surprising, therefore, that they did not want their children to suffer the same deprivation. Why did I use the word 'suffer' in relation to hunger? Perhaps it reveals something of my age, generation and attitude to food?

Not long after the Second World War, my father used to stock our cupboards with tea and sugar packets, just in case we ran short! I used to use the packets to build castles (now I use my unsold books to do the same: to build castles in the air, that is). I remember thinking how lucky we were; we always had a mountain of food in reserve. Could this have been a measure of how much my parents cared for me? Could it have led me to develop a 'Fat Mentality'?

Those lucky enough to eat what they like, when they like, can decide to 'go hungry' if they choose. For them, it might simply be another dieting experience worth trying. If you try it, in imitation of Hindus, Jews, and Moslems when thy fast (as part of their religious observances), you might not find it too difficult.

There are those lucky enough to overcome their phobias of heights, flying, spiders, snakes, and the dark. Some will say, after conquering their fear, 'I don't know why I was so worried!' Those who overcome their dislike of hunger, can also expect some benefits. Fasting can heighten perceptiveness and awareness (spiritual and otherwise). It can also improve energy and give the feeling of fulfillment. Many report 'coming alive', 'feeling sharper', and even feeling healthier with fasting.

There are some who fast for several days, and a few who fast for several weeks without a break. Some who contemplate this might fear weakness and death. Although very unlikely, they must remember to drink plenty of water. While you continue to drink, death would come only after many weeks.

How privileged we are living in the west. Food is available everywhere, and death from starvation is much less likely than winning the lottery!

How do you feel about not eating for 12 or 24 hours?

Does the thought make you feel uncomfortable? Would it make you depressed? Is it definitely not for you? If so, an irrational fear of hunger could perpetuate your weight problem.

It is important for you to know that your inner 'Fat Mentality' is not averse to telling lies, and feeding you with misconceptions about food and hunger. You might hear it say:

- 'What point is there in getting hungry?'

- 'It's unhealthy and uncomfortable to fast, so why do it?'

- 'Life is too short to make yourself miserable.'

- 'Enjoy yourself: you could be dead tomorrow!'

Did your parents ever say these things? If so, your 'Fat Mentality' has followed their lead.

From the moment a baby is born, a mother's role is to keep her child warm, safe, happy and well-fed. As an adult, your 'Fat Mentality" is there to continue the job.

The Fear of Hunger

Every day, millions of people on earth have too little food to satisfy their appetite, yet many of us in the advantaged western world, cannot remember when we last went hungry. Do you, for instance, know anyone who goes hungry? If you don't, you are not alone. Indeed, having succeeded in life, you may have no wish to associate yourself with anyone who knows hunger.

Consider these for a day:
- Try to remember when you were last hungry.
- Is hunger so bad?
- Is hunger to be feared?
- Do you eat to avoid hunger, satisfy your greed, or to prevent unhappiness, loneliness and boredom?

Since many have never known hunger, few will have any reason to fear it. So what makes us eat when we are not hungry? Give it some thought. You must come to understand why it is we do this.

Focused Thinking

How often do you practice focused thinking? Try it! Sit quietly at your desk; on a rock looking out to sea, or over a mountain pass! You might even try sitting cross-legged, eyes closed, beneath a fig tree (otherwise known as a Bodhi, or enlightenment tree, under which the Buddha is thought to have sat). Wherever you choose to do it, have no music playing, and no TV or radio in the background. Go somewhere alone, or at least, somewhere where nobody knows you. Now concentrate.

Ask yourself, 'Is there a question I need to answer?' It might be, 'Why do I eat when I am not hungry?' Go through the possibilities one by one and focus your mind on them. Cross-question yourself. Do I eat when I'm 'bored? Am I unhappy? Am I troubled by stress? Why do I constantly open the fridge door to find food?'

After thinking in a focused way for ten minutes, record your conclusions. Thinking can be exhausting. Don't do it for too long. Your mind will get fatigued and your attention could wander.

A Hunger Experiment

- First, try to eat nothing at all for half-a-day each week, then,
- Try to have only one meal a day.
- Try these on days when other things occupy you.
- Now try them on days when you are unoccupied.
- Try whole days of fasting. One each week to start with.
- Some use alternate fasting days to achieve success.

The point is to get to know what hunger feels like.

When you sense hunger, try drinking a glass of water. The effect is to reduce hunger.

What will you discover?

- You could discover that hunger is not as bad as having a tooth extracted without anaesthetic (or some other pain you have experienced).

- You might learn that being bored or unoccupied, is part of your eating problem.

- You will learn what part thirst, and what part genuine hunger play, in your need to put something in your mouth.

- You will come face to face with your 'Fat Mentality', and notice the voice in your head which says, 'Why bother with this?' 'Why go hungry like this?' 'What's the point in denying yourself like this?' You will learn (if you haven't already) what part the gratification of your 'Fat Mentality' is playing in your need to eat when you are not hungry.

- You may quickly find that eating less will cause constipation. If this is the case, eat more roughage and drink much more water. Drink water until your urine becomes light in colour. If you get dehydrated, you will absorb all the fluid in your large bowel (colon), and cause you to be constipated. At the same time your urine will become darker.

Forgotten Lessons?

Gluttony is one cause of obesity, but not every overweight person is a glutton. We have only recently come to know that our genes and our gut and brain hormones, have more to do with appetite, satiation and weight control than we thought. Various things, other than food, can be the focus of overconsumption (gluttony). Among them are work, exercise, sex, alcohol, smoking and other drugs. The process always involves some loss of self-control, and a strong need for self-gratification. Many lose their ability to say 'NO' to consuming more.

The Appetite Cycle

Without riding the appetite cycle, no diet will work for long.

• When you cut down the quantity of food you eat, your appetite WILL steadily reduce.

• A reduced appetite will lead to a further reduction in your desire for food.

Understanding the appetite cycle will take you where you want to go.

Less Food
Results in:
Less Appetite
Results in:
Less Food
Results in:
Less Appetite

Figure 3: The Food / Appetite Cycle

You Have Two Appetite Reducing Options

1. Introduce fasting days into your week (a quick method) or,

2. Gradually cut down the quantity of food you eat, until you feel comfortable with less food and fewer calories (a slower method).

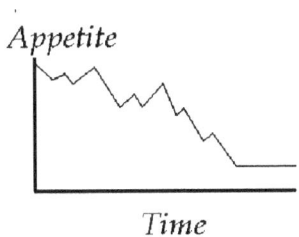

Figure 4: Appetite Progress with Time

The progress of appetite reduction is never smooth. It goes up and down, day by day. Take the long view. Count weeks, not days. If you persist, your appetite will get smaller.

The next step is to learn how much food is sufficient.

Chapter 10

Food Quantity Control.

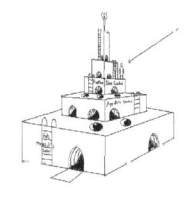

Figure 5: The Dieting Pyramid: Portion Size

If you are overweight, you have eaten more food energy than you need for body maintenance, brain activity and exercise. You have achieved this, either by eating whenever food presents itself (grazing), or by regularly eating portions of food that are larger than you need to maintain the same weight.

Many of us eat far more than we think. (See Brian Wansink. *Mindless Eating: Why We Eat More than We Think*. Bantam, 2006).

War & Rationing

During the Second World War in the UK, food was rationed. Many of the foods that were available, were in short supply. The whole population had no option other than to eat smaller portions.

Shops, supermarkets, bars and restaurants make their profits by selling food and drink. The more attractive their food looks, the more of it they sell. The more we eat, the greater will be the income for these businesses.

No law yet exists that restricts the amount of food sold or the amount we can buy and eat. I suspect that in every country where obesity presents a financial problem, there will be bureaucrats thinking of ways to restrict food– 'in the public interest', of course.

The financial aspects of the food industry, have something in common with the tobacco and alcohol industries. Despite all the negative health implications of both, large profits and tax revenues accrue. Would any government be financially reckless enough to suggest reducing food, tobacco and alcohol sales? This would halt the ever-growing profits made by supermarkets and perhaps reduce the political donations they make.

One Hundred Calories

Do you know how much of each food you like, contains 100 calories? Since the total number of calories needed every day for a weight-reducing diet (for sedentary people) is about 800, it would be helpful to know for instance, what 100 calories of strawberries and cottage cheese both look like. Obviously the portion size (its volume) of strawberries containing 100 calories, will be far greater than for cottage cheese. Three hundred and seventy grams (13 ounces) of strawberries, and ninety nine grams (3.5 ounces) of cottage cheese, both provide one hundred calories. One can eat 3.7 times more strawberry by weight, than cottage cheese, and consume an equal number of calories. Cottage cheese is a much denser food than strawberry, so the volume of cheese needed to get 100 calories will be much smaller than the equivalent portion of strawberries.

Your progress

Just in case any mention of rationing, and reducing the amount of food you eat dispirits you, remember that we all have days when our appetite is strong, and other days when it is weak. We all have days when everything goes well, and others when we will struggle. These latter days are best written off as failures.

It is important to remember:

- No progress ever went smoothly.

- After days and weeks of eating less, you will want to eat even less.

- You will have good days when you find it easy not to eat.

- You will have bad days when it is impossible not to eat a lot.

- On the days you don't feel like eating, resist any additional eating.

- On the days you can't resist extra food, eat minimally (and low carb).

- Persist with eating less food until you feel comfortable with smaller amounts.

- Success at this stage means saying 'YES' to hunger.

- Constantly ask yourself: 'What's so bad about feeling hungry?'

- Don't weigh yourself until you are content with eating less food.

Should you reward yourself twice every week by eating a little of what you fancy? The idea of food as a reward, is pure 'Fat Mentality' thinking. You might do it to begin with, but if you continue, the implication is that you are just waiting for the day when you can revert to a 'Fat Mentality' lifestyle.

Growth Steps

Life progresses in steps. One of the more difficult ones is learning how to control your consumption: the amount of alcohol you consume; the amount of conversation you use to persuade others; how much television and social media you watch, and how much exercise you undertake.

In my experience, most overweight people need to learn much more about food energy intake and energy expenditure. These are lessons they may have failed to learn or have forgotten.

Mentality often changes with age. With age, many become more inclined to self-gratification.

Overweight people who want to lose weight often ask, 'What am I allowed to eat?'. If this is what they ask first, it suggests their failure to ask more important questions. These are, 'How do I conquer the desire to eat unnecessarily?' and 'How do I defeat my Fat Mentality?'

Imagine this:

- You have filled your breakfast bowl with porridge or muesli.

- You then find that half the bowl was sufficient to satisfy you.

What do you do next?

1. Carry on eating until the bowl is empty?
2. Stop eating and throw the rest away?

If your 'Fat Mentality' is in control, you will carry on eating! And how would you justify that to yourself, I wonder?

- 'It's going to be a very long day.'
- 'I might feel faint if I don't eat enough.'
- 'I work better on a full stomach.'
- 'I won't be able to stop for lunch.'

The Third Big Step

This entails being able to:

- Cut down the amount of food and calories you eat.
- On the days you choose to eat less, aim to eat only 800 calories.
- In Appendix 3, there is a list of common foods (in food groups), with the weight of food containing 100 calories of energy. Initially, you can have a free choice of food. Later on, you may have to minimise the foods rich in carbohydrate (sugar and starch). For now,

become acquainted with the food portion sizes that contain 100 Calories.

- Don't change what you eat for the moment, just the amount you eat.
- As the days roll by you should notice that cutting down gets easier.

To eat less, here are the options:

1. Cut down the amount of food you eat by not eating at all during the day. The problem with this is that your craving for food may defeat you.

2. Cut down the amount you eat on several days of the weeks. Eat less, and less often, than inclined to.

3. Work towards cutting down the amount you eat on six out of seven days each week.

4. Always use smaller bowls and plates than you are used to.

5. Eat tiny tasters of low-carb food, throughout the day (like small pieces of cheese, or melon).

If you are a grazer like me, you will find all but the last one difficult. You might have to tell yourself:

'I'll have half a hamburger today, not two double bacon cheeseburgers. I'll limit myself to 10 fries and not eat all I am served.'

Consider having four types of day in every week:

1. Days when you eat less, but eat the type of food you like (at weekends only perhaps).
2. Days when you just eat less, and less often (some weekdays).
3. Semi-fasting days where you eat only once a day.
4. Days when you drink lots of water and eat nothing (reserve for after you have made progress).

Always drink lots of water (one or two litres daily). This will reduce your appetite. Try drinking water instead of eating snacks.

Please excuse so much repetition. Repetition is boring, but there is no better way to learn. So once again, stop and wonder why you should bother to lose weight. Look at yourself naked in the mirror. Do you like what you see? There you will find the answer to the question, 'Why should I bother to lose weight?'

Calorie Counting: AI to the Rescue

There is now an AI-based app, CaloAI®, that purports to analyse the food in a photo. It will provide you with a breakdown of the nutrient and calorie content (amount of carbohydrate, fat and protein). Its reliability and accuracy need to be checked. At least it could help those who do not want to weigh their food or refer to tables of nutrient data.

The Mirror Mantra

- Look at yourself in any mirror large enough to see your whole body.

- Don't hold yourself in or stand tall. Let your tummy hang out!

- You are now going to speak to yourself!

- Repeat what follows, many times.

- Don't just repeat the words aimlessly, put some meaning behind them.

- Repeat, until your urge to eat is less.

Point to yourself in the mirror, and while looking yourself in the eye, say:

'Look at what eating too much has done to you!'

Because you eat when you are not hungry, you have a 'Fat Mentality'.

If you like what you see in the mirror, there is, of course, no need for this.

If you don't like what you see, try repeating the mantras at least twice every week

Another Useful Mantra

'What's so bad about feeling hungry?'

Here are some tips for those finding the going tough.

TIP 1. You have tried smaller portions but find you want more. You see tempting food around you everywhere, and in your fridge. Ask: 'Do I really need to eat this?' 'Am I really hungry enough to eat this?' The answer may be 'Yes', in which case - eat it! If your answer is 'No', recognise this as a critical moment. The food you now want to eat is not food you need to eat. This food will feed your 'Fat Mentality'. You might see it and want it, but not because you need it. When you can say 'No' in this situation, and mean it, you will have made real progress.

TIP 2. Revisit your seven-day diet plan. Assign days to eat eight, 100 Calorie portions. On some days eat some extras. On some days eat fewer. Try alternating liberal eating days with stricter ones. This will test your discipline.

TIP 3. Forget disciplined dieting when you are stressed. If your job is at risk, or your marriage is strained, this is not the time to add another stress.

TIP 4. Never fantasise about food. Never indulge in thinking about food, or relish thoughts of the food you might eat later. Don't mull over the wondrous food you ate last night, or wax lyrical about those delectable chocolates you were given for your birthday. Any undue reverence for food is typical of a 'Fat Mentality', and likely to be part of your problem.

Tip 5: Don't buy the calories. If you don't buy the calories, you are less likely to eat them! Beware of others who want to give them to you instead.

Tip 6: Reject Food: Go to a supermarket, and find the foods you should avoid (see below). Pick up each one, and imagine eating it. Then replace it, saying, 'No. This is not for me'.

Rejecting food can be something to feel proud about. Getting some deserved self-esteem for such disciplined behaviour is important.

Remember that it can be your compulsion that makes you eat, not hunger.

Tip 7: You think you are hungry, but actually you might just be thirsty.

- **First:** drink lots of water or whatever low-calorie fluid you prefer.

- **Second:** Divert your attention away from food by occupying yourself. Walk, talk on the telephone, play the piano, read, garden or do the chores. Now is the time to take piano lessons or to learn Spanish. Anything to take your mind off of the self-gratification demanded by your 'Fat Mentality'.

Shopping Behaviour

- Don't buy sweets, biscuits, or sugar to put in your tea or coffee.

- Don't buy chocolates, bread, cakes, biscuits, buns, cakes and other high-calorie, high-carb snacks.

- Buy low calorie, high fibre products: fruit, vegetables, and low calorie yogurts. Put them on display at home. Hide the cakes and chocolate.

- If food is pre-prepared, and convenient to eat, you are unlikely to know what it contains. So don't buy it. Apart from fruit and vegetables, the convenience food you can grab and eat straight away will usually contain superfluous calories.

- Don't buy food that will quickly gratify you (doughnuts in my case!). You can do without any guilt or a sense of failure.

- If you don't have gratifying foods in your cupboard or fridge, you can go without.

Obesity and food buying habits go together. If you choose food carefully, with its weight-gaining potential in mind, you will have taken an important step towards controlling your body weight. If you don't have fattening, carbohydrate-rich foods in your house, you are less likely to eat them.

One Hundred Calories of Every Food

Hopefully, you have arrived at the point where:

- You have successfully cut down your portion sizes.

- You only eat when you are hungry.

- And eat only at meal-times.

If so, you are ready to learn more about calories and portion sizes. If you are not yet ready, go back to reducing your appetite, with some professional help if necessary (counselling or tablets to suppress your appetite).

Portion Size Training

First you need to know what ½ ounce, 1 ounce, and a pound of food looks like. If you prefer grams, it would be 15, 30, and 500 grams approximately.

Buy a range of different size bowls, from small to large. Put some of your favourite food into each of them, and weigh them. You need to know what 100 grams of food looks like. You can then judge 50 and 200 grams for yourself.

I am not suggesting that you weigh your food each time you eat (unless you have time to spare), just train yourself to recognise different weights of each food. Accuracy is neither necessary nor practicable. All you need is to be approximately correct.

Train your mind. If you are a big eater, you may view 'large' portions as 'small'. If you are anorexic, you may regard 'small' portions as 'large'. Either way, you will need to return to portion size training many times, until you have a true idea of the actual quantities of food on your plate. You need to know how much you are about to eat.

Large or Small Portions? Is it all in the mind?

A three course meal for an anorexic would be an appetizer for a glutton.

Let's look at a few portion sizes containing 100 calories of energy.

Table 1
FOOD
Weight of Food in Grams and ounces containing 100 Calories

Food	Weight
RICE (LONG-GRAIN WHITE)	28 (1 ounce)
RICE KRISPIES	36 (1 ounce)
SALMON (PINK/BAKED)	47 (1½ ounces)
SARDINES	45 (1½ ounces)
SAUSAGES (PORK)	43 (1½ ounces)

You can see that if you eat 1 ounce of white rice, or 1½ ounces of salmon, sardines or sausage, you will be adding 100 calories of energy to your metabolic fuel tank. But what do those amounts look like (their volume), given that a bowl of Rice Krispies weighs so much less than a plate of salmon?

Next you must learn what 100 calories of each food looks like. For this,

- YOU MUST initially weigh your food choices several times, until you can judge the weight of each food that contains 100 calories.

- YOU MUST get someone to test you, and repeat it, until you can recognise the correct amounts.

- YOU MUST succeed at this before you move on. You must learn to judge food weight because you cannot carry weighing scales with you.

- YOU MUST be prepared to re-check yourself every so often; perhaps every month, just to make sure you can judge what weighs 100 grams, and how much of each food contains 100 Calories.

Here are further food choices showing the weight of food that contains 100 calories.

Table 2

PORTIONS OF FOOD CONTAINING 100 CALORIES

FOODS   Grams & Ounces of Food
(Divide grams by 28.4 for ounces)

FOODS	Grams	Ounces
TOFU	137	4.8
POTATO (FLESH & SKIN)	74	2.6
BEANS (BAKED)	119	4.3
RICE (BROWN)	28	1.0
CORN	90	3.2
ASPARAGUS	400	14.1
PEPPER (GREEN)	190	6.8

TURKEY (BREAST ROASTED)	65	2.3
COTTAGE CHEESE	99	3.5
LETTUCE	769	27.5
BRAZIL NUTS	15	0.5
PASTA	35	1.2
CHICKEN (DARK MEAT)	54	1.9
EGG (WHITE)	278	9.9
BEER (one pint)	345	12.3
EGG (YOLK)	29	1.1
CHICKEN BREAST (ROASTED)	65	2.3
WINE (RED)	147	5.3
RICE based cereal	36	1.3
EEL	93	3.3
SHREDDED WHEAT	26	0.9
TOMATO	588	21.0
STRAWBERRY	370	13.2
MAGARINE (POLYUNSAT)	13	0.5
WINE: WHITE	152	5.4
TURKEY (DARK MEAT)	56	2.0
POTATO (SWEET)	115	4.1
BROCCOLI	183	6.5
BANANA	86	3.1
CAULIFLOWER	357	12.8
ONION	278	9.9
TEA	unlimited	unlimited

A Seven-Day Plan

Plan your WEEKLY calorie intake. If you are a sedentary person, and doing less exercise than a manual worker, you must consider eating much less than she or he does.

We all have days when we find self-discipline easy. On other days, our self-discipline can fail us. This makes a seven-day regime the only practicable way to plan your eating. Over a seven-day period, you have time to accept the natural variations in your appetite and your need for satiation.

If you know the sizes (weight or volume) of all the 100 calorie portions of foods you prefer, you next need to know which foods contain the least amount of carbohydrate.

A seven-day, reduced calorie weight loss regime, for an inactive person might provide the following calories (Kilo calories) each day:

- Monday   400 calories
- Tuesday   800 calories
- Wednesday   400 calories
- Thursday   800 calories
- Friday   400 calories
- Saturday   1000 calories
- Sunday   1200 calories

Total weekly calories: 5000 calories

A seven-day weight loss regime for an active person might be:

- Monday   800 calories
- Tuesday   1000 calories
- Wednesday   800 calories
- Thursday   1000 calories
- Friday   800 calories
- Saturday   1000 calories
- Sunday   1200 calories

Total weekly calories: 6600 calories

You could eat eight hundred calories as 8 separate, 100 calorie portions, or as 16 half portions. Many variations are possible. If you enjoy one particular food, have two 100 calorie portions. If you don't like a particular food much, halve its portion size.

Take the time to look at the following list. Again, the weight of food containing 100 calories is show to the right. Choose a few foods you like. Weigh them to get a clear idea of what 100 calories of each food looks like. Make this a regular exercise to re-enforce your skill.

We all know about fishing stories. The size of any fish caught, grows with time. Something similar can happen to your idea of what 100 calories of food looks like. To keep on track, repeat the food weighing exercise at least every month.

See Appendix 3, for a list of 100 calorie portions (shown in food groups). For most packaged food, refer to the calorie and nutrient content shown on the packaging.

The longer list shows that you can eat lots of lettuce (25 ounces or 1½ pounds), tomato (one pound), cauliflower (10 ounces; just over half a pound), and onion (9 ounces; half a pound), and still only eat 100 calories of each. You could also consume 100 calories with half an ounce of Brazil nuts, one third of an ounce of margarine, and one small glass of white wine (approximately one tenth of a bottle).

CHAPTER 11

FOOD CHOICE: CUTTING THE CARBS

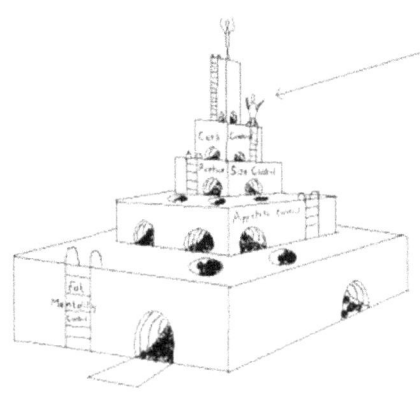

Figure 6: The Dieting Pyramid. Carbohydrate Reduction

The next step, to minimise how much carbohydrate you eat, can prove difficult. That is because a love affair with carbohydrates (carbs) is the commonest cause of weight gain. When people get stuck trying to lose weight, it is often because they eat too much carbohydrate (sugars and starch).

With the amount of food you choose to eat, never use what athletes eat as an example. They can eat piles of pasta and cream cakes (lots of carbs) before heading off to row a race, play rugby or run a marathon. They need lots of available fuel for their exercise, and carbohydrate supplies it. Those who do not exercise, but eat the same amount of carbohydrate, will convert their excess carbohydrate intake into stored body fat.

Food energy is measured in Kilocalories (1000 calories). When burned as body fuel (to power the activity of our body), one gram of carbohydrate yields 3.75Kcals of energy; one gram of protein yields 4Kcals, and one gram of fat yields 9Kcals.

One gram of fat in food contains two and a half times as many calories as one gram of carbohydrate, but to provide us with energy, our bodies burn carbohydrate first.

The focus in this chapter is how to cut down your carbohydrate intake.

Many of us prefer carbohydrate: ice-cream, bread, cakes, sugary snacks, and confectionery. It is possible that some of us even have a genetic predisposition to desire it more than protein. For some it can seem like an addiction. The withdrawal of an addictive substance from an addict, will cause them to become physically ill. The withdrawal of a habit, never causes physical effects. Carbohydrate cannot be regarded as addictive, because its withdrawal will not cause physical effects (pallor, sweaty, low blood pressure, agitation, etc.).

Carbohydrate-rich food contains lots of calories, for immediate use. When we eat too much carbohydrate, the body does not need to burn any of the fat we have stored. The body will store any unused carbohydrate as glycogen (in the liver and muscles) and fat.

The body stores very little carbohydrate as glycogen in muscles and the liver (1.2 kilos); only enough for 20–24 hours of activity. After that, it must burn fat for energy. (Instead of 'burn', I should use 'metabolise'. Nothing in our body actually gets burned).

Some Carbohydrate Science

Barbara O'Neill (www.selfhealthreads.com), in her many social media presentations about health and nutrition, correctly says that the amount of dietary carbohydrate we eat is up to us; namely, it is negotiable. Proteins, fat and vitamins are not negotiable since they are essential for building and repairing the cells in our body and for survival.

There is a direct connection between eating carbs and body weight. To maintain health, the body needs a constant supply of instantly available energy. This is similar in away, to a car engine that needs accessible petrol. Although the energy we need to live comes only from three sources (carbohydrate, protein and fat), the body always draws on carbohydrate first for its energy supply. Only after it has used all the carbohydrate available, will it burn fat. The more carbohydrate we eat, the longer will we hang on to our stored fat and remain overweight.

There are three types of carbohydrate (sugars, starch and fibre). Many, but not all carbohydrates, raise our blood glucose (any carbohydrate that does is called 'glycaemic'). The body uses only glycaemic carbohydrate for energy. Fruit sugar (fructose), does not raise our blood sugar (it is non-glycaemic), and is not a direct source of energy. If we eat fruit containing fructose, it will cause little weight gain. Even though some foods are sweet with fruit sugar (like honeydew melons), they good to eat as part of a low-carb diet. Fruits also contain fibre. Fibre is useful for bowel function and for switching off our appetite.

Only when we have used up the energy stored in the carbs we have eaten, and all that we have stored as glycogen (very little), will we burn fat for energy. During a prolonged famine, when all our fat has been burned for energy, the body will finally start to burn protein (muscle). This happened in the concentration camps of the Second World War, and can still be seen today when disadvantaged, disabled, and refugee camp survivors, eat insufficient food.

How are we to know where (glycaemic) sugar and starch (carbohydrate) are in food? There is no easy way. One way is to note the presence of fleshy white material (starch) in food, such as in bread and potatoes. The white material is starch, a compound made from strings of glycaemic sugars. Sweet foods most often contain glycaemic carbs as glucose and sucrose (table sugar). Those foods containing fruit sugar (fructose) and artificial sweeteners, don't count as calories, because they will not raise our blood sugar.

A typical person going about their business will burn 100 to 150 grams of carbohydrate each day. A completely sedentary person might use only 100 grams; a manual worker could easily burn 200 grams of carbohydrate each day. If we eat more carbohydrate than we burn (on maintaining body functions and exercise), we will gain weight.

In Appendix 5, you will find a list of foods (listed in food groups) that are all OK for the arteries and heart (together with their carb content). In Appendix 4, you will find foods that might make artery disease worse. Their carbohydrate content is also shown. Pre-packaged foods should show the amount of carbohydrate they contain, printed on the packaging.

Look for foods that together provide 50 - 100 grams of carbohydrate per day.

If you want an app to count the number of carbohydrate grams in food, try CaloAI®.

Your yearning for sweets, deserts, bread, cakes and biscuits needs to be overcome. Until then, carbohydrate reduction will remain a problem.

You can give yourself a treat by eating 90% cocoa chocolate (one 100g bar contains only 18 grams of carbohydrate), but eating it could feed your 'Fat Mentality', and maintain your interest in sweet things.

Ryvita® (4.8 grams of carbohydrate per slice) contain much less carbohydrate than a slice of bread (approximately 13 grams of carbohydrate). You could eat quite a few pieces of Ryvita®, each with sliced meat, smoked salmon or cheese (all high protein, low carb), and remain well within your carbohydrate ration for the day.

If you have stopped losing weight, ask: 'How much carbohydrate am I eating?'

The Atkins Diet

Dr. Robert C. Atkin's, of dieting fame, was a New York cardiologist who introduced a now famous low-carb diet (Dr. Atkins New Diet Revolution.1992. Vermilion). His aim was to prepare his obese cardiac patients for heart surgery. His diet minimises the amount of carbohydrate eaten. It maximises the protein and ignores the fat in food. It is very effective, but difficult to keep to long-term. Without addressing a 'Fat Mentality', even this diet is not likely to succeed in the long term.

Eaten long term, the saturated fat allowed in the Atkins diet, could be bad for our arteries. The food choice I advocate, includes those foods that potentially protect our arteries. I have referred to the effect as 'potential', because the benefit of any food said to protect those with a strong inherited tendency to coronary heart disease, remains questionable; the genetic influence to 'fur' arteries can be strong, and not easily negated. I am more certain of the benefit to be gained from avoiding those foods shown in Appendix 4.

In keeping with the Atkins diet, I advocate a low calorie, low carbohydrate diet, made up of foods that are potentially good for the heart and circulation (I first advocated this in my book 'Eat to Your Heart's Content (2005). I could have called it my 'HeartSense' diet).

You can see from Appendix 3, that 100 grams of some cereals (bran flakes, oatmeal, rice-based cereal, cornflakes) can contain one daily ration of carbohydrate. You would have to drink 100 glasses of average white wine (not dessert wine), to get the same amount of carbohydrate. That makes sense, only if you know that most dry white wines, together with most spirits, contain little carbohydrate. Beer contains much more.

A low-carb diet can have several side-effects at first. Those I noticed were:

- A smelly breathe caused by ketones (breakdown products of fat metabolism). See later in this chapter.

- Increased thirst

- Headache

- Feeling less happy, if not a little depressed.

- Constipation (helped by drinking lots of water).

Some facts about food content can be surprising. That is why you might have to re-orientate your ideas about carbohydrate, if you are to achieve permanent weight-loss success. You might need to,

- Learn the truth about portion sizes.
- Learn where the carbohydrates are in food.
- And not stick to what you believe about food without reviewing the facts.

A high protein diet (fish, meat and nuts), with only a little carbohydrate, is more difficult to eat than one rich in carbohydrate. It can be more convenient to eat ready-made bread and cake, than to prepare a steak or salmon cutlet, even though they contain exactly the same number of calories.

Carbohydrates come in a thousand and one, mouth-watering, ready to eat patisseries: from A (Apricot Danish pastries) to Z (Zabaglione Cheesecake).

A Thought Experiment.

- Imagine standing in front of a shop window full of uncooked fish or meat. Would your mouth be watering at the thought of eating them?

- Now imagine standing in front of a shop window full of handmade chocolates and assorted pastries. Is your mouth watering at the thought?

If your mouth waters at the second thought, you prefer carbohydrate-rich food. You could be a carboholic.

If you need help with low-carb recipes, there are many resources. Dr. Atkins' book contains many recipes and an excellent list of the carbs to be found in food. (Dr. Atkins New Diet Revolution.1992. Vermilion)

Many chefs have constructed low-carb dishes. For inspiration, search the internet for Tom Ketteridge's books. Having lost 12 stones in weight himself, he has some sensible advice on how to eat well, while cutting the carbs. See also the Ketogenic Kitchen (2023) by Domini Kemp and Patricia Daly.

The Stone Age Revisited

There may be something to learn from the diet of prehistoric humans. Stone Age man (before 3.300BC) had little access to carbohydrate. There were no bakeries, and certainly no ice cream parlours or shops selling confectionary. Rice, and sugar (sucrose) had yet to appear, but in some areas honey and sugar cane might have been available.

Genetic analysis supports the idea that Wild Einkorn Grass, once mixed with Goat Grass, to produce Wild Emmer Grass (c. 30,000BC). Through gradual hybridization, Spelt, Durum Wheat and Bread Wheat, made their appearance. Around 8,500BC, humans might have first grown wheat and made bread.

Except for some pre-selected individuals, prehistoric human metabolism would not have been adapted to eating carbohydrate. Our present day metabolism is not so well-adapted either. As a result, some who eat lots of bread and sweet things (glycaemic food), increase their chances of affecting their insulin hormone and developing diabetes.

The Ketone Key

If you weight loss gets stuck, the answer to achieving further loss, may be your need for less carbohydrate; the intake of which might have to become negligible.

If you want to know if you are succeeding in this endeavour, there is an easy way. You can test your urine for ketones (derived from burning body fat). From this you can see whether your dietary carbohydrate intake is low enough to need the burning of body fat.

Ketones have a specific smell, and appear in the breath and urine at the same time. Buy yourself some 'Ketostix®'. These dip-stix strips, change colour when ketones are present in your urine. Your pharmacist should have them in stock or can order them for you.

Do ketones help curb our appetite and help with weight loss? The answer is 'yes' to both questions. These effects may be small, but every little helps.

Some Technical Facts about Ketones:

The brain, heart and muscles can use ketones for energy. Ketones can increase the metabolic rate within fat cells, and reduce oxidative stress. They have also been associated with diminishing cancer cell growth (See Bikman, B.T., Fisher Wellman, K.H. in Bibliography).

Low-carb diets and ketosis, both reduce ghrelin production (a gut hormone). The appetite suppression caused, will aid weight loss. When a low-carb diet is found too difficult to tolerate, ketone ester drinks might help. (See Bibliography. Deemer, S.E., Plaisance, E.C).

An unpleasant way to find out if you are making ketones, is to be told that your breath smells. If you eat no carbs with breakfast, and take a ride in a car with a good friend, that friend might smell ketones on your breath and be brave enough to tell you. (There are other causes for an unusually smelly breath. Some need medical or dental attention).

Aim to reduce the carbs in your diet, and keep your urine ketones positive (a state called ketosis). You can use this method to learn which foods are best for weight loss and which are not. After you eat enough carbohydrate containing food, ketones will quickly disappear from your breath and urine. These will be the carbohydrate-containing foods to avoid, if your weight loss has stopped.

That 'Fat Mentality' Again

Zobesians buy all the food they fancy. They buy it, regardless of whether it makes them put on weight. Their ambition in life is to 'take it easy', with no restrictions. That is why they possess many electric devices, including toothbrushes (although many dentists actually advise them). It is far too much effort for them to brush their teeth manually. Self-indulgence interests them most. Self-indulgence makes them feel good about themselves. They have a 'Fat Mentality'.

Fitrakians mostly buy fruit and protein. Unless they are about to begin an athletic challenge, they avoid all the energy-dense food that might make them fat. They see food as fuel; a biological necessity, not a source of comfort and happiness. They hope to be admired for their diligence, hard work, physical fitness and intellectual accomplishments. They are proud of their athletic prowess and do not want to compromise it by eating the wrong food. They have a 'Fit Mentality'.

At heart, are you an Zobesian or a Fitrakian?

Chapter 12

Food for a Healthy Heart and Circulation.

Food 'Goodies' & 'Baddies'

A patient came to me one day as an emergency. He had noticed a tight feeling in his chest, every time he walked. I quickly proved he had angina, with coronary arteries that were severely 'furred' and narrowed. He needed an urgent heart bypass operation.

After his successful operation, he asked me, 'What should I eat? I know that foods containing saturated fat are likely to be bad for me, but which foods might actually be good for my heart arteries?'

The 'furring' of arteries (atherosclerosis) narrows them, and is the cause of angina, heart attacks and some strokes. Finding foods which might benefit the process would answer an important question.

At the time, there was little valid scientific evidence linking artery 'furring', with beneficial food nutrients. What was available, was based only on animal experiments. Some of the research I found, dated back to the early 20th century. I proceeded to identify the experiments which linked the 'furring' process with the nutrients fed to experimental animals.

After animals (usually rats and rabbits) were fed certain vitamins, minerals, amino-acids and fibre, they developed less atheroma ('artery 'furring') See my books for more detail: Eat to Your Heart's Content. The Diet and Lifestyle for a Healthy Heart.(2003), and HeartSense. How to Look after Your Heart.(2006).

After experimental animals (rabbits and rats etc.) were fed saturated fat (animal fat), and trans-fat (formed from frying food), their arteries developed 'furring' (atheroma). Such experiments have not been possible in humans; I had to rely on animal experiments alone for any valid evidence.

Human Artery 'Furring'

Starting when some are in their thirties, arteries can start to develop patches (atheroma plaques) of mixed fat, calcium and scar tissue. This happens progressively over several decades, but probably only in those genetically predisposed to it. The process is called atherosclerosis, but I have referred to it here as 'furring', since it superficially resembles the furring of water pipes in hard water areas.

There are big differences between the furring of water pipes in hard water areas, and the 'furring' of arteries. Water pipes fur when calcium compounds deposit on their inner walls. In arteries, the inner lining (the intima) generates its own cholesterol, calcium and scar tissue. The patches or plaques that form, can grow large enough to block an artery. The process is strongly inherited, and driven more by genes than food. Those who have not inherited adverse genes, are less likely to develop 'furring', whatever they eat. For those patients with evidence of artery 'furring' (ultrasound and CT scanning evidence), I would advise them not to eat much saturated or trans-fat.

Plaques full of fat are the most dangerous (we call them vulnerable plaques); they can split open and attract clot formation. Clot can quickly form and block the blood flow completely. If the blockage is complete, it can cause the death or damage of any tissue that lies beyond. A heart attack occurs when some heart tissue dies (cardiac infarction). When the heart muscle is starved of oxygen, because of a partial coronary artery blockage, angina may occur (this is tightness in the chest felt on effort or with emotion). Angina is a reliable symptom of coronary artery disease (See Appendix 2, for heart symptoms).

Taken together, atherosclerosis and cancer, cause most middle-aged deaths in the western world.

Most foods contain both good and bad nutrients, with a net effect that could determine the presence of absence of the 'furring' process.

After identifying which nutrients are involved in the artery health of animals, my next job was to found where they were in the foods we eat. The British Library had only two books showing the relevant nutrient data for various foods. With this data, I calculated a ratio of good to bad nutrients, and called it the Cardiac Value® of food. Each result allows the comparison of one food (or food group) to another, in terms of its likely benefit to artery health.

Because I found complete nutrient data for a few foods, I could make a start. With my calculated Cardiac Value, it became possible to make some statements about food groups and artery health. Based on chemical nutrient data, offal and shellfish are best of all (packed with beneficial nutrients); ice cream, and dairy products are the worst. Fruit and vegetables contain little of benefit other than fibre, and eating them because we think they will benefit our heart and arteries, has little scientific justification. Fibre can lower blood cholesterol, help to prevent colon cancer and reduce appetite, so it is not without value. For those further interested in the subject, please refer to my books on the subject (see in other works by the author).

My cardiac value results, allow those who want to lose weight and promote their artery health, to optimise their food choices (especially important for those with heart disease and those with a strong family history of angina and heart attacks).

Further research showed that no diet that I constructed, provided every protective substance on a daily basis. Like Dr. Atkins, I also advocate supplementation with minerals (magnesium, manganese, selenium and zinc), B vitamins, and for those who dislike shellfish, the amino acid taurine. You will find my justification for these suggestions in my books.

The Atkins Diet

In his book describing a low-carb diet, Dr. Atkins claimed that his diet, which allows cheese and other fatty foods, had no deleterious effects on the cardiac health of his patients. This may well be the case because atherosclerosis takes a long time to develop. I doubt, however, that he tracked the amount of 'furring' in his patient's arteries, while they followed his diet. I entirely accept that his patients appeared not to suffer any adverse effects.

In keeping with the results of animal experiments on arteries, I advised all my patients to reduce their saturated fat (solid fat like lard), and trans-fat (fried foods) intake. I also advised them to get sufficient amounts of the amino acid taurine (from eating shellfish or supplementation), and the amino acid arginine from eating pumpkin seeds, pine nuts, spirulina, chicken and turkey breast and soy milk.

In my book 'How to become Heart-Smart', I have provided further information about the early detection and management of heart and artery disease. Read it if you have heart disease, or if you have any coronary artery disease in your family.

Other Medical Aspects of Diet

Two cardiac conditions cause 50% of all the middle-aged deaths in the western world. These are high blood pressure (hypertension), and the 'furring' process in arteries (atheroma or atherosclerosis). Cancer causes the remaining 50% of deaths.

High blood pressure is a major cause of strokes, and table salt (sodium) can adversely affect those with high blood pressure (hypertension) who are salt-sensitive. In Appendix 6, the salt content of some foods is listed. In Chapter 18, you will find some further discussion about the relevance of salt in food.

Cholesterol in food is not the problem most people think it is. I have often heard people say, 'I will not eat that, it contains too much cholesterol.' When fed to animals, only 50% of cholesterol is absorbed, and what does reach the circulation, does not necessarily cause 'furring'. It has a bad reputation which is not deserved. Shellfish contains more cholesterol than many foods, although one prawn cocktail contains only 1/400 of an ounce of cholesterol. What is not much appreciated is that shellfish (especially mussels, whelks and oysters) contains many cardio-protective nutrients. Other heart protective foods (which contain more good than bad nutrients), are nuts and seeds (omega 6 & 9 oils), offal (kidney and liver), and oily fish (omega-3 oils).

A strict, low-fat diet can reduce heart attacks by 16%. A Mediterranean (or Lyon) diet, full of protective substances, but not low in fat, reduced the occurrence of second heart attacks by 70% (see de Logeril in bibliography). De Logeril's study suggests that the protective nutrients in food, are effective in preventing heart attacks, even for those who already have artery 'furring' (the reason they had a previous heart attack). Perhaps saturated fat is not as deleterious as we once thought (given the amount of meat and dairy produce in the Lyon diet).

Food Labelling

The major food chains have agreed on food labeling. Some labels now show how much bad stuff (fat, salt and sugar) their food contain. But what about the protective nutrients? Some labels mention 'fibre', but what about the other fifteen 'goodies' I have identified? I understand that no food manufacturer would want their food labelled as 'bad for the heart'; vested interests abound in billion pound industries.

There is an important distinction to be made between 'healthy' food (harmless at best), and food that is actually 'beneficial' or 'protective' (food that might prevent disease).

Can we protect our general health while pursuing a weight-loss diet? What is good for the heart and circulation, is also likely to be good for our general health. Given that heart disease is the number one killer of middle-aged people in the western world, our priority should be to know what is good for our heart and circulation, especially for those who already have heart trouble or have it in their family.

It is not possible to eat a 'good for the heart' diet if you eat too much carbohydrate. Most 'good for the heart' foods contain little carbohydrate, little saturated or trans-fat, and lots of artery protective nutrients.

The Cardiac Value (CV) as a number, indicates how good or bad a food is likely to be for our arteries. Those of neutral cardiac value (neither good nor bad for the circulation) have scores around one (distilled water scores one); those that are beneficial, have scores between one and thirty-one (high scores, like those for offal, indicate the best foods for artery protection). Those bad for the heart, have CV values less than one.

See Appendix 5 for a list of beneficial foods and their CV. By choosing from this low-carb, low calorie list, you can help support your artery and heart health.

Chapter 13

Mantras to Defeat a 'Fat Mentality'

Can you rid yourself of a 'Fat Mentality'? Of course not! It is an innate characteristic, like being short or tall, or having blue or brown eyes. Both a fit and fat mentality co-exist permanently in our minds, albeit expressed to a greater or lesser extent in each of us.

If you are overweight, do you spend most of your time in the 'Fat Mentality' mode? If so, your 'Fat Mentality' is controlling you. It will help you to lose weight if you switch on your 'Fit Mentality', to the detriment of your 'Fat Mentality'.

An awareness of your 'Fat Mentality', and the effect it has on you, is the first step to controlling it.

To change your mindset, try giving yourself a good talking to. This may not be applicable to all, but for those interested, here are a few simple steps to take.

Look at yourself in a full-length mirror

There will be those who only have to look at themselves once in a mirror, to get enough conviction and motivation to take action. The results of over-indulgence and inactivity should be obvious. Yes, I know you know. You don't want to be treated like a child, but **looking,** and *accepting what you see,* are not the same.

Although many try to ignore it, deny it, or hope the image of their body size and shape will quietly go away, what we see in a mirror is usually undeniable. There is one caveat. There are those who perceive a distorted image, like those seen in a fairground hall of mirrors. Some see themselves as extra-large when they are not; some see themselves as extra-small, when they are not. Their perception is distorted.

Some of us dislike being reminded about how we look, although owning up to our image, must precede any possibility of change. If losing weight is a pleasing thought, start by accepting personal responsibility for what you see in your mirror.

If you want to swim, be prepared to get wet; if you want to have a baby, first get pregnant. To lose weight, you must first accept what you see in your mirror. Either accept what you see as your desirable natural beauty, or be prepare yourself for change. It's your choice.

Re-iteration

To change one's mental outlook without some strong motivation and commitment is not possible. Once you have these, a method worth trying employs ***reiteration***: repeating the same message again and again, until its significance sinks in. This is self-indoctrination.

Talk to yourself in front of a full-length mirror.

•

Here's what to do:

- Strip off completely.

- Stand in front of a full-length mirror.

- Don't pull your tummy in; let it hang out!

- Look yourself in the eye.

- Point an accusing finger at yourself.

- While pointing your index finger at yourself, accuse yourself of these things:

"YOU have a Fat Mentality – YOU eat, even when you are not hungry."

"YOU have a Fat Mentality – YOU cannot say 'NO' to food."

"YOU have a Fat Mentality – YOU eat regardless of being overweight."

"YOU have a Fat Mentality – YOU eat more than you need."

"YOU have a Fat Mentality – YOU too easily give in to self-gratification."

"YOU have a Fat Mentality – YOU take it easy more than YOU should."

Delete or add your own accusations, and recite them regularly.

Repeat them after rising in the morning, and before going to bed at night. Do it every day. You might slowly come to accept the truth about your mentality.

Reinforcement Exercises

These exercises could help you deal with your dominant 'Fat Mentality', in other ways.

Walk around a supermarket with no intention to buy anything. One by one, pick up the foods you like; the ones you know contain too many calories and too much carbohydrate. Handle them and then reject them, while thinking 'this is not for me'.

Look at 50, 100, 200, and 500 grams of porridge in separate containers. Practice guessing the different weights.

Collect together some high calorie, high carb foods. Put them together on a table and recite the mantra: 'Eat these and get fat'.

Isn't all this rather naïve? Yes, it is! Isn't this just New Age, touchy-feely twaddle? Yes, it is! Despite that, these simple procedures can be effective for some stuck in a 'Fat Mentality' mode.

If you have owned up to your dominant mentality, you are ready to learn more about appetite control, portion choice and exercise.

If you cannot accept the truth about your mentality and its command over you, and still want to lose weight, you may need further help. Ask a sensible, intimate and trusted friend, to tell you what they see. If you don't want to do that, consider professional help to overcome what could be a distorted perception of your body image.

Chapter 14

A 'Fit Mentality' is not for Everyone.

There are obvious differences in mentality between those who are overweight and those who are slim; those who are fit and those who are unfit. If you have failed to lose weight several times, you might need to learn a little. Even if your reaction to having a 'Fat Mentality' has been, 'Yeah, Yeah! I get it!', do you really understand how different that is to having a 'Fit Mentality'?

The differences between successful and unsuccessful business people; a good cook and a professional chef; athletes and the indolent; the average golfer and a professional golfer, can be small but all are crucial.

Consider this. If you are a yachtsman, an adventurous rower or an airline pilot, whose destination is New York, rather than in Washington DC (about 230 miles apart), all you need do is to head out of the English Channel, less than one degree further to the north. Small changes like this in starting conditions can lead to quite different end results. You will remember the butterfly effect mentioned in chapter seven. Adding a small amount of exercise, or slightly reducing your carbohydrate intake, can effect a significant weight change later on.

It is important to set your time parameters. Small initial changes will take time to have any effect. Large initial changes, like repeated daily fasting, will get results more quickly.

If changing your body weight and fitness has defeated you, it might be instructive to understand why those with a 'Fit Mentality', have no weight problem. Is it simply because they eat less and exercise more? Not entirely, because athletes undertaking heavy training, can eat five times more calories per day than their unfit rivals, yet still not put on weight. They burn off the extra energy with intense exercise, and burn more energy after exercise, and while sitting still.

A key question for those concerned with weight loss is:

How am I to eat less and exercise more, day after day, without having a 'Fit Mentality'?

Let's return to doughnuts. Imagine them stuffed with jam; doughy, delectable and irresistible. The thought of jam draws you closer. The satiation of your desire is just one bite away. The mission embodied in every doughnut is to seduce you and satiate you. Some will be to buy a pack of six and eat them all.

How is it that some people can eat just one half of a doughnut, and refuse the rest? If you are one of them, would your explanation be:

'I like doughnuts, but I don't need to eat them all.'

or

'I've just been to the gym. I'll eat one to restore my energy'

or

'No thanks. I don't want one. I only eat when I'm hungry!'

or

'I work hard physically so I can afford to eat some!'

If any of these apply to you, you have a 'Fit Mentality'.

Let's go back to the 'Fit / Fat Mentality' questionnaire. Return to Chapter seven, and check your 'Fit Mentality' score.

If you scored 3 for a 'Fit Mentality', you might switch into a 'Fat Mentality' occasionally, but should remain in control. If you scored over three, you hardly need weight control advice, unless you want to grow your muscles or become aerobically fitter. If you scored three or less for a 'Fit Mentality', your 'Fat Mentality' probably dominates you, and losing weight could be a struggle.

By now you should have a fair idea of what I mean by a 'Fit Mentality'. There are, however, different aspects to it, other than eating, keeping slim and remaining fit. Let's take a look at this mentality as it applies to food, and other important aspects of life.

These are all 'Fit Mentality' characteristics:

Applied to Food:

- Food does not easily seduce you.
- You are happy to eat your own meals, and not those of others as well.
- You mostly eat, only when you are hungry.
- You are not a grazer.
- You eat large portions when training.
- You have little reverence for food, drink and restaurants.
- You prefer protein: steaks and fish, rather than carbs.
- You use food as fuel, not as a reward.
- You might say, 'Why waste time eating, when I could exercise.'
- You are very specific about what to eat and how much.
- You avoid 'all-inclusive', pre-paid buffets.

Applied to Activity

- You love exercise and physical challenges.
- You believe in exercising your body and mind.
- You have little need for comfort.
- You have little need for self-gratification.
- Achievement gives you a lot of pleasure.
- Your motto: work hard and play hard.
- You prefer economy to waste.

Applied to Money:

- Your business ethic is to maximise profit with hard work.
- Your ultimate aim is success and the improved self-esteem it brings.
- You delegate only to free up time for work on other enterprises.
- Continual money accumulation is nice, but not your primary aim in life.
- Winning counts.
- You see small amounts of food as sufficient.
- You are not proud of your purchasing power.

So you see, a 'Fit Mentality' doesn't always mean being thin and fit, but your attitudes and associated behaviour patterns make it more likely.

Sometimes, a 'Fit Mentality' can be exaggerated. One of my young female patients had a sculptured, muscular body. She ran several marathons every year and attended a gym morning and evening. She ate like a sparrow, determined not to add one ounce of body fat. Her regime started during a failing relationship. She came to believe that her future happiness, and sense of fulfillment, depended on her physical fitness and lack of body fat. Another of her misguided beliefs was that happiness and contentment depend entirely on financial wealth.

There is little doubt that most of those with a 'Fit Mentality' are likely to be physically fit. There is no reason to believe, however, that they are healthier mentally. Set in their ways, selfish, and impervious to the views of others, some can be a complete 'pain in the arse' when trying to convince others of the benefits of their disciplined way of life.

Those with no sporting interest at all, have only to be around amateur athletes, talking endlessly about their game and how they won or lost it, to be bored senseless. Their heady mixture of ego and insecurity, is easily dampened. If you have the courage, ask them, 'Who cares who wins? After all, it's only a game!' Are there not more important things in life than winning games and races? Like winning the weight loss game, perhaps.

None of this is relevant to professional athletes. They need to make their living by winning contests. Professional sport is big business, where 'getting to the top', can make athletes very rich. Many a boxer and footballer, with an uncompromising drive to win, has progressed from rags to riches using their sporting skills.

Whether you chose sport, business, home life or a professional career to achieve success, happiness and self-esteem count. What you choose to eat doesn't much matter much if you are happy with the outcome. You can choose to be:

- Fit, slim and happy, or,
- Unfit, overweight and happy.

These goals are matters of choice. Both are yours to choose.

The growing prevalence of obesity will cost governments lots of money, but that is a price democratic societies must pay for contented citizens with some freedom of choice. Does signing up to the rules of a society, and paying taxes for mutual benefit, give us the right to be the weight we want? I think so, but if you don't agree, perhaps you think we should all be slim. There would be an advantage. We could keep more of the tax collected and spend any surplus on armaments.

Chapter 15

Eating Habits, Eating Styles & Culture

Eating satisfies hunger for a while, but to a variable extent. That's biology. What it may not satisfy is the desire to eat. That's psychology.

The motivation to eat regardless of hunger is one aspect of a 'Fat Mentality'. The compulsion to have more, regardless of need, is greed. If this describes you, don't feel bad. You can blame a well-developed survival trait. Every human trait and instinct once had a biological purpose; a good appetite for both food and life, can foster the survival of the fittest.

Our adult drives to eat and procreate are primal, and responsible for the survival of humanity. Because they have an evolutionary purpose, any attempt to overcome them will be resisted. When we try to reduce our food intake, we challenge the strong evolutionary forces that have enabled us to survive and develop over millions of years.

In many cultures today, thin people are more often seen as beautiful than fat people. In pre-historic times, slim fit guys must have had a disadvantage. While keeping on the move to avoid dangers, they would have needed to eat frequently. Without a handy corner shop nearby, the constant challenge would have been to find the next meal. They ate to live. Although the slim fit guys could move fast (burning lots of energy in the process), they might have lacked the stamina of their larger, more ponderous companions (with larger stores of energy for long-distance foraging).

The 'Fat Mentality' drives us to eat through:

- Plain greed. A straightforward compulsion to eat, with no reason other than, 'I feel like it, so I'll eat it!'

- Uncontrolled self-gratification and self-indulgence, and

- Well-formed habits.

Appetite alone drives those with a 'Fit Mentality' to eat. Lust for food occurs only when they are hungry.

It's my metabolism doc!

The reasons people give for their obesity, or their inability to lose weight, can read like a list of excuses. Those who gave me their reasons for failing to lose weight, did not always see them as excuses. Some of their reasons were:

- 'I feel ill when I don't eat.' This might mean: there is a good medical reason for eating, too dangerous to ignore (like a peptic ulcer in the stomach).

- 'I haven't time to go for a walk or go to the gym.' This means: My priorities are different.

- 'I've always had a slow metabolism.' This means: I haven't had it tested, but this assumption allows me to deny what it is I eat and how much.

Just occasionally, a defective metabolism is the cause of obesity or being underweight. I considered these in chapter 7.

Eating Habits and Culture

Our style of eating can only partly explain why some of us become overweight.

The way we eat may depend on our cultural background. My father was born in 1904; a time in history when many in the UK were lucky to have much food. He never worried about overeating.

In many middle-eastern and eastern cultures, they will put many dishes of food on the table at the same time. This allows the greedy to serve themselves liberally; they can eat large amounts of food without others noticing. This contrasts with the western tradition of serving food in consecutive courses (derived from the Court of King Louis XIV, in the lead-up to the French Revolution).

Some trim, weight-conscious people, are to be seen pushing small amounts of food around their plate. This is the opposite of food gorging. The idea is to convince others they eat normally, when actually they eat sparingly.

For those who can afford to eat in restaurants regularly, many eating styles can be seen. Self-control and self-gratification are both on display.

Some cultures have no conception of fixed mealtimes, and no rules about what to eat at different times. Others have strict rules for both that are accepted without question. Those who live with such rules may not consider them odd, until they experience the habits of foreigners.

An American cardiologist colleague of mine, New Yorker Dr. P.K., once told me he preferred to take his holidays in Acapulco. In the hotel he chose, there were three restaurants. One served breakfast on a 24-hour basis; the others served either lunch or dinner around the clock. This meant it didn't matter what time he awoke; he could get breakfast, lunch or dinner, anytime he wanted!

In Russia, they thought me eccentric when I suggested having tea and cake between three and four in the afternoon (as is traditional in Great Britain). The reaction was, 'Surely, you can drink tea and eat cake at any time?'

For most of their lives, my parents (my father was an Edwardian) had no notion of taking afternoon tea outside of the three to four pm time slot, or of eating dinner (which many working people called 'tea') later than 6pm.

Many still believe that it is 'bad to eat late'. They believe that eating before going to bed will somehow cause a greater weight gain (relative to eating early). There are two reasons why this might be true. First, our body clock slows down at night, and so does the rate we burn calories. Second, the relative inactivity of non-dreaming sleep burns less energy. When considering weight gain, these factors become insignificant compared to the total number of calories we can eat during the day. Eat a lot and do no exercise, and you will put on weight. Since it can take an hour or so of physical activity to burn off 500 Calories (Kcals), it might be better not to eat the calories.

At 10.30pm., in July, in a small deserted Taverna in Athens, I asked a server if it was too late to eat. He laughed and told me it was actually too early! 'The musicians and our customers arrive soon', he said. By 11.30pm the place was buzzing with customers, eating and enjoying the Bouzouki music.

As a young Englishman, I would never have eaten porridge at any time other than breakfast. In the UK, there are still some who only eat fish on Fridays. Exposure to culturally accepted habits from birth is the reason. Other, less indoctrinated nations, eat whatever they like, whenever they like.

When I was young, I was often told that I ate too fast. I have always eaten fast. When I was a junior doctor, I had the reputation of being able to eat a 3-course lunch in five minutes. I could then quickly return to my ward or Casualty (now A & E) to see patients.

From an early age, my mother told me that I would get 'indigestion' if I continued to eat fast. She was right about most things, but I have never had indigestion! Seventy-years later, I am still the first to finish my meal when eating with others. Perhaps this is one aspect of my own 'Fat Mentality'.

Why do we eat what we eat?

In the not so distant past, many British children felt guilty, or were made to feel guilty, if they didn't eat everything put on their plate. After all, there were many starving people in the world. How finishing the food on our plates could have helped starving children in Africa, still mystifies to me. This is, of course, a psychological 'cop-out'. It could have caused some to overindulge, and others to have left the table feeling guilty about their non-compliance.

In restaurants, servers have sometimes asked me why I left some food. I have always been too polite to ask why I should eat all the food given to me, when the chef did not know the extent of my appetite.

Many people choose restaurants based on value for money. Many want large amounts of food for a reasonable price. In choosing an 'eat-as-much-as-you-like' restaurant, many will easily satisfy their 'Fat Mentality'.

Some people eat to live. Some live to eat. For those unable to imagine living to eat, I would advise them to spend a weekend at La Reserve de Beaulieu, near Monaco, on the Cote d'Azur!

'A gourmet is just a glutton with brains.'
Philip Haberman Jr. (1961) Vogue Magazine,

Some eat to get enough energy for their work. Many now work so they can afford to eat. Such factors predetermine, not only what we eat and how much, but also why we eat. Not only will our weight be affected, but sometimes our contentment, happiness, diabetic control, blood pressure and heart disease risk.

Eating and Social Class

It is now anachronistic to mention it, but only those of a certain social class once knew the correct use of fish and pastry forks, and the glasses from which sherry, hock, champagne or claret should be drunk. Those old enough to remember Fanny and Johnny Craddock on TV, will remember how strict they were about compliance with such rules. Except in certain refined social circles, these rules have mostly disappeared.

Nowadays, almost anything goes. Eating etiquette and dress code, have almost disappeared. The once strict rules of eating behaviour have given way to eating at all times, whether we are working, walking or talking.

Modern standards of behaviour allow for eating and speaking at the same time (almost every feature film has an example of it). Also acceptable now is eating in the street and on public transport. When I was a child, these would have been seen as serious transgression of social etiquette. For no reason I can justify, they still make me react with disapproval. In a completely unconscious way, those who eat while talking or on public transport, show no respect for what were once fixed rules. Why should they? In certain places in the UK, however, old-fashioned, class-defining eating etiquette can still be observed.

Non-verbal behaviour is a valuable source of information. What we humans eat, and how we eat it, can reveal a lot about our social background. In the UK, our social status, culture background, and educational influences are still on display as we eat.

Soon after they met in 1931, my father invited my mother to the Lyon's Corner House at Marble Arch, London, for afternoon tea. As a shy person from a poor background in Bethnal Green, East London, my mother was apprehensive. She had never dreamed of entering such a prestigious establishment. To my father's dismay, the situation made her anxious. It took him one hour to coax her inside. She worried that her table manners would give her (social class) away. She needn't have worried. My father found her lack of pretentiousness an endearing feature.

Another manifestation of her social insecurity arose during the Second World War. My mother feared that while King George VI and Queen Elizabeth were visiting the East End of London during the Blitz, they might make an impromptu visit to her house. She would not have wanted to appear lacking in table manners. She needn't have worried. My mother shared some natural gifts with the Queen: grace, charm, politeness, and elegance, all of which would have made such a visit a joy for both of them. The Royal couple never did visit her.

Many cultural differences exist around the world. Indians, eating 'correctly' (according to their cultural norms), would once have eaten only with their right hand, never with their left (thought to be unclean).

The amount of food people eat is interesting to observe. At one end of a wide spectrum, there are those with a lust for large amounts of fat and carbs (binge-eating disorder) - the gluttons; they see only large amounts of food as acceptable. At the other end of the spectrum, are those who eat very little – the anorexics. They will see small amounts of food as more than sufficient.

Eating Disorders

When farmers feed their animals, there will often be a rush to the trough. Some humans act similarly. Although they may try, it will be difficult for them to speak with a full mouth. Forget having a conversation with them while they are eating. Some will nod instead of talk; some will even resent any interference as they feed.

You would be wrong to think that all those who engage in such eating behaviour are overweight.

They can be grossly overweight or slim. Both can eat meals, two or three times larger than the average person. The overweight ones might graze hourly, the fit ones will eat only when they are hungry.

Both can eat with the focus and determination of a person worried about an impending famine.

A physically fit doctor and friend of mine, would give a dirty look to anyone who attempted a conversation with him while he was eating. He ate with a sense of time urgency, always in attack mode. Few of his friends dared to interrupt him. He never became overweight, perhaps because he did weight-training every day.

Big eaters often take large amounts of food with each forkful, so limiting the number of journeys needed from their plate to their mouth. This may be necessary for those in a hurry, but just as often, it is their usual eating style. Their attitude to eating suggests that they are racing against the clock.

In 1974, Rosenman and Friedman named this time-urgent behaviour, Type A behaviour, and found that it associated with a doubling of heart attack risk.

Among food gorgers there is a sub-class of eaters with an Olympian outlook to eating. They are both competitive and egotistical. They will eat mounds of food as an act of bravado or as an attention seeking strategy. Their message is, 'Watch me. I'll show you how to eat!' Should others in their company leave their food, they will offer to eat that as well. They are proud of their super-normal appetite and are ready to inform others of the benefits. Their sense of how much food is healthy, has clearly become distorted. Some are athletes who seem never to put on weight.

Intractable binge-eating, as a recognised medical condition, is best referred to doctors specialising in eating disorders.

Grazing

Grazing describes regular or constant eating as a way of life. Grazers are not necessarily passionate about food; more often than not, they eat because they are bored. Between main meals they will eat nibbles, snacks, mini-meals, morning coffee, afternoon tea, and late-night snacks. Grazing is a common characteristic of overweight people, although fit people who spend lots of energy, are also liable to it.

For many, conversation is an essential accompaniment to eating. Eating and conversation can combine to be one of the great pleasures of life. For those with a 'Fit Mentality', the social interaction can be more important than the food served.

Food Avoidance

For those trying to lose weight, there is something to learn from food avoiders. During a meal with others, they will spend half their time surreptitiously moving food around their plate. They will try to spear the smallest amounts of food on their fork, and then proceed to chew it at length. Only 5% of their mouth will be full at any one time. They can easily manage continuous conversation while eating. Their main aim is to maintain a low body weight by measured self-deprivation. Some bring grace and elegance to eating, and make it an art form. Because they are so concerned about their weight, some will attend slimming clinics to get slimming pills.

Some have a distorted view of both their body weight (which they may not see as underweight), and the amount of food they eat (which they see as normal when it is too little). Some are anorexic, some are bulimic; both can benefit from medical intervention.

Be on the lookout for those with anorexia. You could save a life if you spot one in your circle of friends. They will rarely allow more than a minimal amount of food on their plate; a fear of weight gain inhibits their eating. Fearing obesity, anorexics may suffer severe food deprivation. Devoid of fat under their skin, and both anxious and withdrawn, they may still see themselves as fat, even when their ribs and the bones in their face protrude. This is a sight still to be seen in some refugee camps, and reminds me of concentration camp scenes shown in newsreels at the end of the Second World War.

Anorexia nervosa has one of the highest mortality rates of any psychiatric condition (5% mortality within four years of diagnosis), with bulimia occurring in 25-50% of cases. Both can be more dangerous than obesity. The causes are complicated, often involving family relationships.

Both should receive specialist care in an eating disorders unit.

The Nit-Pickers

Fussy eaters often have an obsessive trait. They will typically have a strict dietary regime in place, which may not include eating between meals. They treat food with respect, measuring the amounts of food they eat, and the amounts they serve to others. One delicate bite of a wholegrain sandwich, or one precise bite of banana followed by a small gulp of carefully chosen spring water, will satisfy them (they are obviously rich enough to pay 1000 times more for each glass of spa water than for tap water).

Some eat in a precise order. As environmentally conscious beings, their crumbs and banana skins will usually be disposed of in an environment friendly, 'green' manner. Because their weight can be one of their neurotic concerns, a few are at risk of becoming anorexic or bulimic.

Undercover Agents

Some overweight people dislike being seen to eat excessively. In the company of others, they will try to eat modestly; some will even attempt to create the impression of being food avoiders. From their appearance they are obviously not, so who do they think they are fooling? They must hate the thought of being called a glutton, and may be trying their best not to eat too much.

While working, some chefs eat continuously. After all, they need to taste the food they serve. Some will hide food and alcohol, and consume it while no-one is looking. Like drug addicts, they can adopt several secret strategies to conceal their habit. All those who eat and drink secretly have a problem.

What I have written so far about food, can apply to alcohol. The focus may be different, but the behaviour is the same. Some big drinkers and alcoholics will quickly drink a few pints of beer, or several bottles of wine, and follow on with a few quick-fire shots of spirit. Some are weekend binge drinkers, others are equivalent to food grazers, and drink continuously throughout the day.

Food and alcohol buying habits are pertinent to obesity. If we choose food carefully, with its weight-gaining potential in mind, we will have taken an important step towards controlling our body weight. If we don't buy fattening food and alcohol, we are less likely to consume it.

PART FOUR

ARE WE WHAT WE EAT?

Chapter 16

How to Exercise More

Humans needed strength, physical endurance, and a resilient nature to survive a prehistoric life. The most intelligent and knowledgeable hunter gatherers would have survived better when they knew how, when and where, to capture their food.

Surviving modern life no longer requires the same physicality; a few now manage successful businesses using their mobile telephone. Many now delegated all the physical jobs to others. This has become the direction of travel for citizens living in developed societies. Exercise as a hobby has slowly replaced exercise at work. We can rely on robots using AI to make many more of us sedentary and liable to obesity.

Exercise is important for health. Even walking as a healthy hobby, reduces the prospect of chronic disease (see Ford and Caperson, 2012). It helps to strengthen our bones and reduces the prevalence of the cardiovascular diseases responsible for half of all middle-age deaths in the western world.(Warburton et al. 2006; Löllgen 2013).

Being selected for exercise ability must once have favoured human evolution. Now all we need to do is park outside a shop. The average bodyweight in many nations is growing steadily, partly because of the growing indolence of the lives many now lead.

We can easily measure the numbers of calories in food. The calories used in various forms of exercise is not so easily measured. The commonest exercise undertaken in gymnasia is treadmill walking and running. This will burn about one hundred calories in 20 minutes (depending on the speed and elevation of the treadmill). Some will think it better not to have eaten that slice of toast for breakfast! If exercise did not lower medical risk, we would have no need to promote it to those seeking weight loss. Advice to eat less would suffice, because the metabolism our body uses a lot of energy without exercise.

Our resting energy expenditure is approximately one Kcal per Kg of body weight, per hour. A 70Kg person will burn 70Kcals per hour, or 1680Kcals every 24 hours.

As BMI increases, more calories per minute are burned. While doing 30 minutes of exercise, a person weighing 185 pounds, will burn 30 to 50% more energy than a slim person weighing 128 pounds (doing the same exercise). In 30 minutes, an overweight person weighing 185 pounds will spend 125 Kcals doing weight training; 159Kcals walking at 3.5mph; 252Kcals swimming; 231 Kcals doing low-impact aerobics; 294 Kcals on a stationary bicycle and rowing; 125Kcals with slow ballroom dancing; 147Kcals playing golf; 294Kcals playing football; 420Kcals cycling at 15mph, while gardening will use 189 Kcals. (See Harvard Health Publishing, 2021 in Bibliography).

So which type of exercise might be best for healthy weight loss? Athletic, normal weight people, will undertake whichever exercise or sport satisfies their competitive spirit; the overweight will need to choose less intense exercises (less oxygen per minute burned). Because they allow frequent breaks, the exercises that best suit overweight people are walking, swimming and resistance exercises (like weight training where the key to burning calories is the time spent with weights under tension. See: Christopher B Scott. Appl Physiol Nutr Metab. 2012 Apr.)

A recent meta-analysis showed, that of all forms of exercise, vigorous exercise reduces mortality most (Samitz et al. 2011).

Another survey confirmed that overweight people (average BMI = 28), naturally do less daily exercise, than those of ideal weight (average BMI = 22). In this survey of subjects aged between 21 and 39 years, fit males used 995 Kcal. per day doing moderately vigorous exercise. Fit females used 568Kcal. per day. Obese males spent 714 Kcal./day, and obese females 406 Kcal. per day. (see Clemens et al. 2015). Coping with high-intensity endurance training will be difficult for adults with a BMI over 24 Kgs/m^2.

Obese people have been shown to overestimate their physical activity. It might help if they were aware of this. (See Lichtman, SW, et al. 1992, in Bibliography).

The first exercise for anyone overweight to consider is distance walking. There is an obvious health gain to performing any form of exercise, even walking. One can walk almost anywhere, although it is more pleasant to walk by the sea than down a busy main road, choked with traffic. To maintain interest, consider driving to somewhere where it is more enjoyable to walk. Because walking allows frequent breaks, it is an ideal starter exercise for the overweight.

First choose a one half mile or one mile circuit (1.5 Km) from home, and walk it slowly with as many breaks as you need. Gradually speed up and decrease the time to complete the same distance. Use several different circuits to relieve boredom. As your confidence grows complete more circuits at a faster pace.

Here are some commonly asked questions about exercise, posed by overweight people.

Q. *I want to lose weight. Should I lose weight before I start exercising?*
A. It depends on your weight. If you find it difficult to walk without getting breathless, lose some weight first. Only consider more strenuous exercise when you can walk at a normal speed without breathlessness. If you cannot talk while walking with a partner, your priority must be to lose weight first.

Q. *Is it OK to take breaks during exercise?*
A. Whatever your level of fitness, this is good practice. It is especially important if you are overweight. Try interval training. Walk as fast as you can until you get breathless, even if it is only a few meters. Stop, or walk very slow, until you regain your breath. Start walking

again, fast enough to get breathless, then stop again. Repeat this stop start method, as many times as you can before feeling fatigued. It is important to get breathless, but not so breathless you feel distressed. You will become fitter, only by causing breathlessness. After repeating this daily walking method, you might find you can walk further and further as the weeks pass (it takes much more than a few days to notice any difference).

If you carry on with this regime for long enough, walking will no longer be enough to make you breathless. You will then need to jog or run. You will need to be moderately fit before you can run without breathlessness.

Q. *What other forms of exercise are appropriate for an overweight person?*

A. Swimming and light weight training are best. Whatever exercise you choose, try to make yourself a little breathless, and then take a break.

Search YouTube for beginner weight training exercises. If you do not want to go to a gymnasium, you will need to get two light dumbbells and a bar for attaching weights (as your strength improves). As you progress, you will need more equipment. Joining a gymnasium then becomes more practicable.

When training, use weights that strain you a little, but not too much. Whichever training exercise you choose, choose weights that allow you to complete ten repetitions (reps) of the exercise. At the end of ten repetitions, you should feel some strain. If you cannot complete all ten reps, the weight you are using is too heavy; if doing ten reps feels too easy, add more weight.

In this way you can progress around ten different machines in a gym, doing only ten reps at each station. This should not take long enough for you to get bored.

Q. *Is it enough to do short bouts of exercise that make me breathless?*
A. Yes it is. This is because it will stimulate your body to improve its aerobic ability (oxygen fuel efficiency). The body is no ordinary mechanical machine; it improves (if healthy) in response to demand.

If you exercise in the way I have described, you will gradually achieve the equivalent of replacing your car engine. You will gradually replace your engine with a more powerful one. Your small one litre metabolic engine will gradually grow into a three litre engine. This is not an overnight process. It can take between twelve and eighteen months, depending on how regularly you exercise.

Q. *Should I exercise every day?*
A. It is good to walk every day, but only do interval training or weight training, three times each week to start with. Only add days of exercise when you have the energy for it. If you do more than your body finds acceptable, muscle and joint pains could hold you back. This applies as much to top athletes as to the unfit. In the early phases

of taking up exercise, wait for any muscle pains to subside, before exercising again.

There may come a time when you can work through the muscle pains you get. Doing exercise can relieve them. To get to that stage will take some time. Adaptation to exercise will happen more quickly for those who were once athletically fit.

Whatever exercise you undertake, avoid dehydration. Dehydration will make you feel thirsty, and your urine darker. Drink enough to keep your urine light in colour.

With weight training, you will grow muscle. Muscle is heavier than fat, so with muscle growth, you may not lose weight quickly. Your shape will change, and your muscles should become firmer.

Chapter 17

Are You Disciplined Enough for Eating Less & Exercising More?

How Disciplined Are You?

I have a little bad news. It will take some discipline to follow any dieting and exercise regime successfully, especially if losing a lot of weight is the aim. Only the disciplined can hope to reverse the strong desire to eat regardless of hunger, and an unwillingness to exercise in the long term.

It might help to know just how disciplined you can be, and how likely you are to succeed, whatever weight loss regime you choose to follow.

A simple questionnaire follows. Fill it in as a test of how disciplined you can be, and how likely you are to lose weight and keep your chosen weight.

Respond in a spontaneous way. Give little thought to your answers. Give no thought to why the questions are being asked.

Each question has four answers to choose from. Choose the answer which best reflects YOUR behaviour as it is NOW.

Taking the test can help you with some insight.

There are no right or wrong answers.

Question 1: Are you always 'on time'?

A: Always.
B: Most often.
C: Not usually
D: I am never on time.

Ring Your answer: Is it: A B C or D ?

Question 2: Imagine you have been called to serve in the Army. Are you likely to....

A: Find a regimented life unacceptable.
B: Hate every minute
C: Try my best
D: Look forward to it.

Ring Your answer: Is it: A B C or D ?

Question 3: You must learn to speak a foreign language in one year. Are you likely to...

A: Refuse
B: Give up when it gets too difficult.
C: Set yourself a study plan.
D: Try harder, the more difficult it gets.

Ring Your answer: Is it: A B C or D ?

Question 4: With difficult tasks, do you....

A: Immediately ask for help.
B: Ask for help when necessary.
C: Know you can do it if you try.
D: Never try. Just get someone else to do it!

Ring Your answer: Is it: A B C or D ?

Question 5: Did you always try hard at school?

A: Yes.
B: Mostly.
C: Only if pushed
D: Never.

Ring Your answer: Is it: A B C or D ?

NOW SCORE YOUR ANSWERS

Question 1: For 'A' score 20; 'B' score 15; 'C' score 5; 'D' score 10.

Question 2: For 'A' score 10; 'B' score 5; 'C' score 15; 'D' score 20.

Question 3: For 'A' score 5; 'B' score 10; 'C' score 15; 'D' score 20.

Question 4: For 'A' score 10; 'B' score 15; 'C' score 20; 'D' score 5.

Question 5: For 'A' score 20; 'B' score 15; 'C' score 10; 'D' score 5.

NOW TOTAL YOUR SCORES, AND AWARD YOURSELF A
GRADE FOR
SELF-DISCIPLINE

Those with a total score of 75 and over, have excellent Grade 'A' discipline.

Those with a score between 50 and 74, have Grade 'B' discipline. Good but not the best.

Those with a score between 30 and 49 have Grade 'C' discipline. Less than desirable.

Those with a total score under 30 have Grade 'D' discipline. Far less than desirable.

What do the results say about you?

Those with Grade 'C' and 'D' discipline are going to find dieting regimes difficult. They must consider improving their personal discipline. Only by taking on a personal ambition or challenge, might this improve. You must then stick with it, and accomplish it. Once you are aware of what discipline you are capable of, you should be ready to try weight loss again.

Those with Grade 'A' and 'B' discipline will not find losing weight much of a challenge. They are most likely to succeed.

Self-discipline is hard for some and easy for others. Have you ever overcome a challenge by being self-disciplined? Did it feel good? If it did, you will need to revisit that feeling. Become an advocate for self-discipline and be proud of your achievements.

1949 saw the introduction of National Service in peacetime. I am old enough to remember a few Teddy Boys of the 1950s, being called for duty. A neighbour of mine, had a son who was a seasoned Teddy Boy. He had a hairstyle referred to as 'duck's arse'. His dress was bizarre, and out of keeping with the formal dress code of the time. His smoking, and lounging around behaviour, drew him a lot of attention. I well remember the time when his call-up papers came for National Service. He was summoned to join His Majesty's Armed Forces.

I remember seeing him when he returned on leave. He was wearing an army uniform, and his hair had been closely cropped. What surprised me most, was the change in his demeanour. Instead of slouching, he carried himself with dignity, as if standing to attention. He stood tall and wore his uniform with pride. Some would have said that the army had 'knocked him into shape'. He had found some self-esteem in being disciplined and was proud of what he had become.

Those patients I saw, who had once served in the armed forces, all had something different about their bearing. They carried themselves differently to others; almost as if they were ready to stand to attention and salute at any moment. These men usually dressed meticulously and all had highly polished shoes. In tight-knit fighting groups, discipline is essential. In the armed forces, nobody survives long unless they conform. That takes discipline. Surprisingly, it can last a lifetime.

Now give some thought to your own life story. When and how often, did you use your self-discipline to overcome challenges? If you need to, picture your successful accomplishments to boost your confidence.

Chapter 18

Weight Loss: Truth & Myth. The Appliance of Science.

We usually get closer to the truth by overcoming our biases, and by trying not to fool ourselves. The truth, like any sculpture, has different aspects that depend on our viewpoint. When several scientists study the same subject and get different results, it can be a problem to decide which one best represents the truth. The skill needed to decide this, lies at the frontier between science and art. Both judgement and interpretation are art forms.

In order to get to the truth, and to dispel myths, doctors mostly rely on evidence from experiments that compare one large randomly chosen group to another.

Myths may not represent the truth, but lessons can be learned from some of them. Here are a few myths and truths about weight loss and exercise.

The only way to lose weight is to exercise.

Diet and exercise need to be combined to get the best results. One can lose weight with a low-calorie diet alone, but exercise will help to burn some calories. The problem with exercise is much time it takes, and how much is needed. One hour in a gym for an overweight person might burn 250 Kcal. To avoid eating a 250 kcal hamburger, is simpler.

My advice is to lose some weight first with a low-carb diet, then exercise.

One cannot deny the considerable long-term health benefits of exercise.

Cutting out certain foods is crucial to weight loss.

This is true. Cutting down the carbohydrate content of the food we eat helps most. However, as long as the total number of calories eaten daily is less than 800 or 1000 (for sedentary people), most people will lose weight.

The body will burn fat more readily with a low-carb diet, simply because carbohydrate is first in line for our energy supply. With only a little carbohydrate in our diet, we will quickly use our stored carbohydrate, and then start to burn body fat (and body weight) for energy. If the low-carb foods we choose also contain less fat, weight loss will be more rapid.

Can ketone drinks help?

This is true. Ketone ester drinks, which mimic the ketogenic effect of a low-carb diet, can help with appetite and metabolism.

Gluten-free diets help weight loss.

There is no evidence that a gluten-free diet aids weight loss. Food without wheat germ (containing gluten) will not always contain less carbohydrate. Unfortunately, gluten-free products are often full of sugar and fat to help mimic the flavour and texture of the products they replace.

People with celiac disease and gluten intolerance, have reason to avoid gluten (wheat protein) in food. When they avoid gluten, they will usually put on weight.

Juicing is a good for weight loss.

If the juice made contains few calories, then it could help weight loss. The problem is that some nutrient content may be missing. Protein, fat and fibre, are all necessary ingredients for a balanced diet. Juicing can reduce the amount of fibre taken, but will preserve the sugar and vitamin content. In addition, the absence of chewing can affect the digestive enzymes we produce and the satisfaction we get from chewing.

Small changes in diet help most.

Researchers (Cornell University's Food and Brand Lab.) found that those who lost most weight in the long term, were those who made small, consistent changes, to the way they eat on most days.

They found the best tips for weight loss were:

• Keep worktops clear of all foods except the 'healthy' (low-carb) ones.

• Always serve a portion of food onto a plate. Never eat directly from a packet.

• Eat something hot for breakfast within the first hour of waking up.

• Have something small to eat every three to four hours.

• Slow down your eating. Put down your knife, fork or spoon, between bites.

(See: Cornell Chronicle (December 20, 2012). Small changes at least 25 days a month prompt weight loss.)

What your weighing scales show counts most.

Our weight can change depending on how much salt and fluid we retain. If you must weigh yourself, do it in the morning after emptying your bladder and bowels. Do not weigh yourself every day. Once every week is enough, although daily weighing can motivate some people.

A change of shape, rather than actual weight loss per se, is what many want. With weight-training exercise, muscles grow. This can cause an initial weight gain. With the loss of body fat and muscle growth, most people will change shape and look healthier. Combining diet with exercise helps.

All calories are equal.

To physicists they are, but other considerations concern biologists.

What is a calorie? Those who studied physics at school, will know that one calorie of energy will raise the temperature of one gram of water, by one degree centigrade. Because this amount of energy is small, we measure food and body energy in units of one thousand calories, namely in kilocalories (Kcal).

The number of calories in food is one thing, how the body uses them is another. Imagine a car with three fuel tanks. The first to be used contains only carbohydrate. It has a very low capacity. When that is empty, the fat containing tank next supplies the fuel. Its capacity is very large, and never gets completely full. When that is empty, the protein filled fuel tank supplies the fuel as a last resort. Because the carbohydrate fuel tank is so small, any excess of carbohydrate eaten, will have to be stored as fat.

The rate at which we burn calories varies with our fitness level. The muscles of athletes burn more calories sitting in a chair, than those who are unfit. They also burn more in the hours after exercising (Excess Post Oxygen Consumption or EPOC). This extra energy use helps weight loss. One can transform an overweight, unfit body into an athletic one, but a lot of commitment will be needed.

Fasting works

If you eat fewer calories over a long period (long-term fasting), your weight-loss progress can stop.

Our body responds to starvation by making alterations in hormone production (from the bowel and elsewhere). The body switches into survival mode. The hormones produced make us eat more, and slow our metabolism. This conserves our reserves of fat.

It is OK to fast for one or two days per week, but alternate these days with normal eating.

Eating late is a cause of weight gain.

There are studies that show eating before bedtime, will put on more weight than eating during the day. For anyone who is overweight, the effect is too small to worry about.

Because eating food can be soporific, some who eat late will get to sleep more easily. I know this, because I am one of them. There are sometimes other, less pleasant reactions to eating late, like vivid dreaming and disturbed sleep. This type of sleep (Rapid Eye Movement, or REM) burns more energy than peaceful sleep.

Aerobic exercise ('cardio') is better than weight training (resistance exercise) for weight loss.

This issue concerns athletes more than unfit and overweight people. Because our metabolism becomes adapted to running and other aerobic exercise, any weight loss benefit will reach a limit. Growing muscle with weight training may be more effective for some in the long-term.

High-intensity interval training will aid weight loss, but unfit, overweight people, will find it difficult to perform.

You can reduce weight in specific areas of your body (like your tummy).

This is a myth.

It is not possible to burn the fat from one specific area of the body. Fat is metabolised from every area at once.

If you want to reveal your six-pack to the world, lose weight overall.

Alcohol is not fattening.

Oh, yes, it is! It's the calories that count. There is a further consideration though. Beers usually contain carbohydrate, spirits do not. In theory, drinking beer should reduce weight loss, and can cause weight gain. The type of alcohol drunk is a matter of choice. The best advice is to stop alcohol altogether (too many calories) or give preference to spirits or dry wines.

In early Victorian times, they made beer with added milk. Milk stout (or stout porter) contained lots of calories, and given to nursing mothers, labourers, and underweight sick people, helped to energise them and put on weight. In 1946, the label 'milk stout' was banned in the UK, although adding lactose sugar is still used to give beer a creamy head.

We need willpower to lose weight.

Self-discipline and will-power are important when overcoming challenges. For some people, however, no amount of willpower will help them overcome the influence of their genes and gut hormones.

A lack of self-discipline and will-power typifies a 'Fat Mentality', and is a common cause of repeated weight loss failure. Overcoming this mentality, and constructing an acceptable eating and exercise regime as part of a healthy lifestyle, are all essential for those who want to lose weight.

Conditioning our mind to avoid temptation (as discussed in Chapter 10) using various mantras, while exposing ourselves to foods that are best resisted, can help to reduce the influence of a 'Fat Mentality'. It is tempting providence for some, but keeping biscuits on display and rejecting them, can pay dividends. This simple provocative technique might boost self-control, but takes will-power.

The sugar in fruit causes weight gain.

Fruit sugar (fructose) does not raise blood sugar (it is not glycaemic), therefore it is OK to eat on a low-carb diet. Some fruits, like honeydew melon, contain mostly fruit sugar. Others like grapes, contain a lot of glycaemic sugar, together with fructose. Eating lots of grapes is, therefore, not OK on a low-carb diet.

The quantity we eat is important. One can eat much more melon by weight than grapes, because of the glycaemic sugar in grapes.

It is accepted that eating a few pieces of fruit each day is healthy. Fruits are good sources of fibre and vitamin C. Fibre is important because it can reduce appetite, and slow the absorption of glycaemic sugars. Fruits mostly contain water, and apart from fibre, contain nothing cardioprotective. Those fruits that contain red, blue and purple compounds (flavonoids or polyphenols), may be an exception.

Some fruits, tea and red wine, contain polyphenols. Blueberries, red grapes, red wine, plums, strawberries, green tea, tomatoes, red peppers, blackcurrants, pomegranate and many other naturally coloured foods contain them.

Polyphenols are antioxidant, and for many valid scientific reasons, may benefit artery 'furring' (See Ziółkiewicz, A. et al. in Bibliography). However, nobody has yet demonstrated a direct beneficial effect of a prolonged polyphenol-rich diet on the 'furring' process.

Having breakfast is important for weight loss.

Some say they think better on an empty stomach. Mid-morning hunger can be a problem for those who don't eat breakfast. One good thing about the Atkins, low-carb diet, is that it supports having a filling bacon and egg breakfast, but without toast, pancakes, jam or maple syrup (full of carbs like starch and sugar).

Diet foods and drinks help us lose weight.

This is true for sugar-free drinks, but then you might have concerns about the artificial sweeteners they contain. Look for the carbohydrate content on the label. Remember that a low-carb diet allows for only 50 to 100 grams of carbohydrate each day.

Sweet drinks (however sweetened), will help maintain your interest in sweet foods, and the weight gain some can cause.

If you keep to whole foods, and stay away from processed foods, you will avoid a long list of additives.

Too much salt is bad.

Some salt (sodium chloride) in our diet, is essential to life. All our bodily fluids are salty, and need replenishing. Because we all need a regular supply of salt, only too little or too much can be bad for us.

That salt is bad for us is now a common assumption. Almost every patient I had, accepted that salt was bad for them, and had cut their intake down as much as they could (adding none to the food on their plate). As a result, some became prone to faintness (low blood pressure) because sufficient salt is essential for the maintenance of blood pressure.

We need a pinch of salt each day (approximately 1.6 grams [1600mgs]: the RNI, or recommended daily nutrition intake), although in hot weather, we will need a lot more if we sweat. The same necessity for a sufficient intake, also applies to other chemical salts composed of potassium, calcium, magnesium, manganese and iodine.

Professor Graham MacGregor, Professor of Cardiovascular Medicine at the Centre for Public Health & Policy, once worked in the same department of medicine at Charing Cross Hospital, as me. To the best of my knowledge, it was he who initiated the idea of dietary salt as a cause for high blood pressure. He also promoted the idea that adding salt to food was ill-advised for those with hypertension. At this time, I was managing cases who fainted regularly because of their low blood pressure. As an important sub-group, they can benefit from extra dietary salt (In bibliography see: Yuan, Ma, et al. with MacGregor, G.A.)

Salt can cause high blood pressure and raise blood pressure from low levels to normal levels (for those who are liable to faint). It can also raise high blood pressure further. A high dietary salt intake is also associated with obesity. It is for these reasons that salt has its bad name.

Water retention in the menopause, and several other medical conditions, occurs because some hormones cause salt retention. Because salt and water go together, whenever we retain salt, water accompanies it. Because water weighs a lot (try carrying a bucket of water), water retention can cause considerable weight gain.

Doctors can prescribe a diuretic to relieve salt and water retention. A diuretic will make the kidneys excrete salt and water. In cases of high blood pressure, a diuretic alone can sometimes work well (when salt retention is a factor). Diuretics do not help in all cases. Some doctors prescribe diuretics to help people lose weight. If they work, it is because of water loss, not fat loss.

Look for the sodium content on food packet labels.

If 100 units of salt is the healthy daily requirement (=1600mgs), aim to consume less than 100 units of sodium salt each day (unless you have a persistently low blood pressure). In Table 3, you will find an alphabetic list of the foods containing most salt.

Because one portion of anchovy (85 grams) contains 209 salt units, one portion contains twice the daily salt requirement (100 units). The fact that baked beans (100grams) contain 33 salt units, means that a 100 gram portion contains one third of our total daily salt requirement.

All the other foods I reviewed, contained less than 20 units of sodium salt. See Appendix 6 for a fuller list.

Will weight loss make me a different person?

At one end of the reality spectrum, there are those who are overweight, happy, healthy, and poor; at the other end, there are some who are slim, fit, rich and miserable.

Don't put too much faith in weight loss as a cure for depression or a solution for your relationship problems. Weight loss can improve your self-esteem, but then so will achieving any desired goal. Lose weight for yourself, and you might not get disappointed. If you lose weight to comply with the wishes of others, you risk more than disappointment. You risk resentment, losing friends and more.

Although corporate political minds are now focusing their efforts on relieving national obesity, one embarrassing political fact remains. The fact is, poverty is much more dangerous than obesity.

Although most years of life are lost because of smoking (4.8 years) and diabetes (3.9 years), the years lost from obesity (0.7 years), are far fewer than those lost from a low socio-economic status (2.1 years less)(see Brogan, Caroline, in Bibliography).

A strong association exists between wealth, bodyweight, food availability and survival (mortality and morbidity). Although the health divide between the rich and poor, has for millennia been one of the biggest social problems facing humanity, too little has been done about it. Like a poisoned chalice, many politicians will actively ignore it. It's a lot easier to justify their political position by saying, 'That's the way it has always been!'

Here are some unpalatable medico-political facts, that are mostly kept quiet.

- The poor are three to five times more likely than the rich to die from cancer or heart disease.

- The mortality of an overweight group of people can be 40% higher than a matched group of normal weight people (it can be 2.5 times more for those defined as obese).
- The poor and less well educated, are more often overweight than the rich and well educated.
- Excessively overweight people (BMI > 40) are more likely to have diabetes, heart disease and joint problems.

There is one good thing to say about fortune, fitness and weight loss. They are all topics that sell books. 'How to get rich', 'How to get Fit' and 'How to lose weight', can become best-sellers.

Table 3
Foods Containing Most Sodium Salt

Food	Food Weight grams	salt units 100 = Daily Requirement
Anchovy	85	209
Beans Baked	100	33
Bran Flakes	50	25
Camembert	62	23
Cheeseburger	400	170
Cornflakes	100	63
Crab (steamed)	118	31
Eel	140	56
Hamburger	250	97
Lima Beans	78	20
Lobster	145	30
Oatmeal	240	83
Sausages (Pork)	100	75
Tuna	150	30

Chapter 19

Self Help and Medical Help

Many who have found weight loss difficult need more information about weight control and the medical help available. They need to know how best to control their appetite; where the calories are in food; how much carbohydrate there is in food, and which low-carb, low calorie food, is best for their arteries and heart.

Some need psychological help to understand what has driven them to eat too much and not to exercise enough. Depression in all its forms can be a cause, sometimes with a long history. Depression can present in different ways, with many losing their interest in life while trying to cope with bereavement and other stresses.

Cognitive therapy and antidepressant medications can help many aspects of depression. Between counselling, group therapy and psychiatry, help is at hand, but it needs to be tailored to the individual.

What follows is a review of self-help, and the medical help available from physicians and surgeons.

Self Help

Some appetite suppressants are available in health-food shops. There are also food preparations that can help weight loss by affecting bowel bacteria (the gut biome). A brief description of them follows.

Spiralina

This was once thought to be blue-green algae. After doing some genetic studies, spiralina has been reclassified as a bacterium with several species. It has been used as a food supplement by astronauts.

Taking one to two grams per day for twelve weeks on a low-calorie diet, has been shown to reduce body weight by an extra 3-4 Kgs (6 – 9 pounds). It affects fatty tissue directly and improves insulin sensitivity. This improves how well the body handles sugar and fat (it counters the metabolic syndrome). See Chaouachi, M. et al. in Bibliography.

Garcinia cambogia

Garcinia cambogia extract, and Garcinia cambogia containing products, are some of the most popular dietary supplements currently marketed for weight loss. In 2014, $60 billion was spent in the USA on weight loss products. This was one of them. It is an extract of an Indian and South Asian fruit that resembles a tomato.

Its safety is in dispute because of potential liver damage.

Green Tea Extract

This is safe, and has been shown to achieve a modest weight loss (1 to 1.5Kgs over 3 months). It affects the gut hormone ghrelin. (see Chen et al. in Bibliography).

Hoodia

Hoodia gordonii extract reduces appetite over several hours. It is used traditionally by the African Khoi-San tribes to control their hunger while foraging for food. The plant is known as Bushman's Hat; a leafless spiny succulent plant, found in the Kalahari Desert and throughout Angola, Botswana, South Africa, and Namibia.

It works by affecting ghrelin and GLP-1 gut hormones when taken in daily doses of 100 – 150mgs. (see Jain et al. in Bibliography).

Prebiotics, Probiotics, Synbiotics & the Gut Biome

Gut bacteria can have far reaching effects on body weight, bowel inflammation, and even brain activity.

Eating indigestible fibre helps bowel function and bacterial activity. In the colon, fibre is broken down into fatty acids which also influence bowel inflammation. Fibre containing foods are called prebiotic foods. They include whole grains and oats; fruits like apples, bananas, peaches and watermelons; legumes like red kidney beans; green vegetables, asparagus, artichokes, as well as onions, soya products and chickpeas. Probiotics are foods, or food preparations, that contain beneficial bacteria (Lactobacillus acidophilus is one of them). Synbiotics are those foods and preparations that contain both.

The bacterial population of the human bowel, taken as a whole, is called the gut biome (or microbiota). The total number of bacteria in the human bowel exceeds 1014, (10 with 14 noughts following it). Sometimes pathogenic (harmful) bacteria can take hold and affect health. Research supports the suggestion that these bacteria and the chemicals they produce, can influence our health, metabolism and weight control.

There are many different types of bacteria present in the bowel (more than 1000 species). The bacteria present and their diversity, varies between individuals. Some are associated with weight gain, others with weight loss. This is because some bacteria can produce fatty acids - a source of energy that can be absorbed and stored.

The chemicals produced by some bacteria can reduce appetite and increase satiation. Some bacterial probiotics, and synbiotics, can alter the secretion of gut hormones.

In mice, probiotics (containing Lactobacillus acidophilus, and Bifidobacterium longum) have been shown to slow weight gain. Other bacteria have been associated with obesity (Firmicutes).

In an experiment on mice, a high carbohydrate diet promoted the growth of bacteria associated with weight gain.

Exercise promotes the growth of bacteria associated with weight loss. The same researchers suggested that a mixture containing Lactobacillus acidophilus, can help with weight loss.

For those further interested in this research, see Aoun, A. et al. in the Bibliography.

Medical Help

Diuretics

Diuretics are drugs which remove salt and water from the body.

That too much salt in the diet is dangerous, is a view commonly held by members of the public. Although sodium salt is an essential constituent of all our bodily fluids, too little or too much could be fatal. We need to get the right amount of salt each day from our food (see Appendix 6). Each day, we need approximately one pinch of salt (1.6 grams is the recommended nutrition intake), although in hot weather we will need more as we sweat.

The intake of many other minerals is also essential daily. They include potassium salts (bananas and citrus fruits), calcium (dairy produce, green vegetables, beans, lentils) and magnesium (in all the above plus nuts and seeds). A balanced diet will also contain iodine, selenium, zinc and every vitamin.

Doctors have been concerned for decades about the effect of salt on high blood pressure. It can raise blood pressure from low levels to normal (when people are liable to faint), and raise high blood pressure to an even higher level. It is for this reason salt has a bad name.

Various hormones cause salt and water retention in the menopause. Because water is heavy (try carrying a bucket of water), these hormones are sometimes responsible for weight gain.

When this occurs, doctors can prescribe a diuretic. A diuretic will make the kidneys excrete salt and water together. In high blood pressure, a diuretic can sometimes work as the only treatment (when salt retention is a factor). Diuretics can be completely ineffective in other cases.

To help people lose weight, some doctors prescribe diuretics. They can work for a short while, but only because water, not fat, is lost.

Stimulant Drugs

Many doctors are against using stimulant appetite suppressants. Used appropriately, they can work well, but are habit-forming. They can also give patients the extra energy they need for exercise. They can act as a euphoriant (I am not sure if they are directly antidepressant).

In the right dose, they have few side-effects, but can cause insomnia and palpitation. They switch off the normal physiological processes of hunger and satiation. I have known many patients use them wisely, but others who relied on them for energy and a feeling of well-being. Any proof that they adversely affect the heart is lacking. The two most commonly used have been phentermine and diethylpropion (Tenuate dospan). Both are amphetamine based. Another is benzphetamine. Phendimetrazine stimulates the brain to reduce appetite, but is not amphetamine-based.

Dexfluramine and Aminorex were both withdrawn, after safety issues were reported. They were once used to stimulate the brain and reduce appetite.

I did not know that amphetamines had a history of being used as antidepressants, until a highly intelligent, retired Hungarian lady (Eva W.), told me that her father and grandfather had taken them for depression, before the Second World War. Both were depressive characters and the amphetamine they took for decades, improved their quality of life.

Eva had been under the care of psychiatrists for decades. The standard antidepressants they gave her did not improve her. In fact, they slowed her down, and made her morose. Given her family history, I decided to give her a short, experimental trial, of dexamphetamine sulphate (5mgs). What followed was truly remarkable. She became rejuvenated. She moved house and made a new life for herself.

I could not continue the experiment. Local drug enforcement officers arrived at my practice, cautioned me, and stopped me prescribing them. The patient who had known severe depression, and had seen no benefit from any antidepressant, had stockpiled some. She must have decided that a depressed life was not for her. Soon after being told that she could not receive any further dexamphetamine tablets, she used her stockpile of antidepressants to commit suicide.

Patients can die at the hands of dutiful, ruled-based bureaucrats, who lack clinical experience and perspective. I do not know what their reaction was after I informed them of Eva's death. I guess they would have fallen back on the Nuremberg defence: 'I was only following instructions!'

I have told this tale before, and I will continue to tell it in Eva's name. It was my first brush with inane medical bureaucracy, and I intend not to let them forget the responsibility they assume for patients.

One concerning aspect about the use of stimulant appetite suppressants, is their use by those who are already slim, and want to stay that way. An obsession with weight, is sometimes a requirement of their job. Some are addicts and need psychological help. Many are manipulative, and can easily dupe doctors into giving them appetite suppressant prescriptions, even when their BMI doesn't warrant it. Many doctors in the past have made their fortune selling these drugs. In the UK, that practice has now largely disappeared.

Mimicking a Gut Hormone (GLP-1 agonist)

Semaglutide (Wegovy® and Ozempic®) is an injectable weight-loss medication for obese adults. It mimics the naturally occurring gut hormone GLP-1.

By mimicking GLP-1 (it is a GLP-1 receptor agonist), semaglutide increases the feeling of fullness, and reduces food intake, hunger and food craving.

After one year and four months, some trials showed patients losing between 10% and 16% of their body weight. Those on an innocent placebo, lost only 5% during the same period.

To put this into perspective, a person weighing 16 stones (102 Kilos), can expect to weigh between 13.4 stones (85 kilos) and 14.4 stones (91 kilos) after one year and four months of being given semaglutide. Patients lost between 1.6 and 2.4 stones (10 to 15 kilos)(European Medicines Agency. See in bibliography).

Semaglutide is very effective, but with side effects experienced by one in ten people. Among them are nausea, vomiting diarrhoea, constipation, abdominal pains, and muscle weakness (long-term). Although all patients should be supervised, both Wegovy® and Ozempic® are available on the internet.

WARNING. There is an active trade, promoted on social media, for cheaper injections of semaglutide. Some are not what they claim to be, and contain insulin. Because it rapidly lowers blood glucose, it is dangerous to inject without medical supervision. All medicinal drugs obtained without a prescription, are best avoided.

Semaglutide is used for adults with a BMI of 30 kg/m² and over (obesity). It can be prescribed for those with a BMI of at least 27 kg/m², if they have a weight-related health problem (diabetes, high blood pressure, abnormal blood lipids, shortness of breath, sleep apnoea, a history of heart attack, stroke or blood vessel problems).

Adolescents can use it from 12 years of age, as long as their BMI warrants it.

Liraglutide

Liraglutide is sold under the brand names Victoza® and Saxenda®. It is an anti-diabetic medication that stimulates pancreatic insulin production. It is used to treat type-2 diabetes and chronic obesity. It is similar to semaglutide.

Setmelanotide

Setmelanotide is a medication for the treatment of severe obesity, caused by genetic disorders (deficiency of proopiomelanocortin [POMC], proprotein convertase subtilisin/kexin type 1 [PCSK1], or leptin receptor [LEPR]). It acts on the melanocortin-4 receptor (MC4R).

Orlistat

Lipase is an enzyme that dissolves fat in the bowel. Orlistat is a lipase inhibitor. When taken by mouth, it reduces fat absorption by the intestines. Fat calories remain in the bowel, and are lost in fatty stools. (See Suyog et al. in the Bibliography).

Compared to a placebo, orlistat (120mgs three times a day before meals) produced a significant weight loss in eighty patients

(BMI>30) ($p<0.05$). After 24-weeks, they lost an average of 4.65kgs versus a average loss of 2.5kgs by the placebo group.

Sibutramine

The EU and USA have both suspended sibutramine. It is a 5HT and nor-epinephrine reuptake inhibitor (an antidepressant). A 4–6% weight loss after one year of treatment occurred while taking it, but adverse cardiovascular effects, like slight increases in blood-pressure and heart rate affected some. These effects caused it to be suspended.

Belvic

Belvic (lorcaserin) is an appetite reducing medication that acts in the brain. It acts on 5-HT2C receptors in the central nervous system (CNS), particularly those in the hypothalamus. It reduces appetite.

DNP (2-4 dinitrophenol)

This is no longer available in the UK., having recently been reclassified as a poison (October 2023). It was once an ingredient in an explosive mixture used in the First World War.

Many would regard DNP as the perfect slimming solution: weight loss without dieting. It does this by causing the body to burn fat and carbohydrate. Many died using it (less than 100 worldwide), because of excess body heat generation. When this occurs, the resulting high temperature and sweating is irreversible. There is no antidote. In the long-term, it also caused cataract formation.

The Russian military once used it to keep their soldiers warm during severe Russian winters.

Heptral (S-Adenosyl-L-methionine or ademetionine)

Heptral 400mgs (once or twice daily before food), helps to maintain the metabolism of proteins, fats, lipids and hormones. It can improve the function of the liver, especially when it is fatty (also for those who drink alcohol heavily).

Some of its other effects are surprising. It promotes weight loss, improves mood, and makes many feel energetic (it can make sleeping difficult if taken after 2pm). Take this medicine only when suggested by a doctor, and in the dose prescribed.

Psychological Interventions

The aim is to support the cognitive, emotional and social aspects of weight control. There are several types of talking therapy, each with a different focus. Those with psychological difficulties, and a problem with weight loss, need to choose the most appropriate support.

CBT (Cognitive Behavioural Therapy) aims to change patterns of thinking and behaviour. CBT can help with anxiety and depression, and also help to change any unhealthy attitudes to food and exercise.

Motivational Interviewing can help with ambivalent attitudes and boost motivation related to food, exercise and weight loss. It can help with commitment by enhancing personal feedback and reflection.

Mindfulness Interventions. Self-awareness and self-compassion are encouraged. It can help to deal with the judgment of others.

Acceptance and Commitment Therapy aims to create psychological flexibility, empowering the commitment to achieve goals.

Social Support Interventions recognise the value of social networks. Group and family therapy is encouraged, as is networking with like-minded people. For some, a sense of belonging can enhance motivation, and help with a long-term commitment to weight loss. Many companies offer this type of support together with expensive dieting food products.

Surgical Intervention

Strict weight criteria apply to those who want surgical intervention to help them lose weight. Only those with a BMI greater than 35, are eligible. Some who are desperate for surgery, have been known to carry rocks in their pockets, to boost their weight and eligibility.

Those who are overweight, and have a fit mentality, might be the most suitable for surgery. They may have a metabolism problem (hormonal or genetic). These are people who eat all the right things, do regular exercise, and still have a resistant weight problem. Because they have a 'Fit Mentality', their chances of maintaining a lower weight after surgery, are better than those who have failed to suppress their 'Fat Mentality'.

The simplest surgical intervention is the introduction of a **gastric balloon (intragastric balloon)**. A saline filled bag is positioned in the stomach. It induces a feeling of satiation, and lessens the desire for food.

Gastric banding employs key-hole surgery (laparoscopy) to place a ligature around the upper part of the stomach. It results in a feeling of fullness, after small amounts of food. When necessary, the band can be tightened or loosened.

Open surgery is required to reduce the size of the stomach. A **sleeve gastrectomy** removes a longitudinal strip of stomach and makes the stomach more like a sleeve than a bag. Small amounts of food achieve satiation, and long-term weight loss is common.

Gastric bypass. This entails sewing the small intestine into a pouch made from the upper stomach. No food reaches the stomach or duodenum, and the gastric juices normally found there, are no longer available for digestion. Fewer nutrients are absorbed, and a substantial weight loss can result. Because less dietary fat is absorbed (malabsorption), patients experience permanently fatty stools.

This is an operation reserved from the obese (BMI > 35), and those grossly overweight (BMI > 40), who need help to control their breathing (sleep apnoea), diabetes, and heart disease. Infertility provides another indication.

A duodenal switch operation combines a sleeve gastrectomy with a gastric bypass. The stomach empties into a short, small intestinal segment, not the duodenum. The duodenum delivers its normal digestive juices, but food does not reach the same bit of small intestine as it normally would. Only those with a BMI > 35 are usually considered.

Chapter 20

Rounding – Up. Slimming Down.

This book contains all the most pertinent information for those seeking weight loss. It is for those who want to lose weight, to choose which bits of information suit them best. This information can put them on the right path, but only they can walk the path. The path is not an easy one to tread; a few steep inclines will be encountered along the way.

All those who want to lose weight must come to understand and control their 'Fat Mentality'. Determination, commitment and stamina are all required. Some will have to try several times before reaching their goal. I have no wish to be pessimistic, but on this path, action needs to replace false hope.

Those with a minor weight problem (Class I (overweight, and a BMI between 25 and 30), a weak 'Fat Mentality', and some discipline, should simply increase their walking and cut down their daily carbohydrate intake of food. They should easily lose weight in the course of a few months.

By simply cutting out one food completely, like bread and cake, effective weight loss can be achieved. For some it is that simple.

Those who are more overweight (BMI>30), and have a strong 'Fat Mentality', will probably need all the other interventions described in this book, although cutting down on carbohydrate-rich food is the single most effective intervention.

Exercise is essential for general health, but is not time efficient; exercise burns too few calories per minute. Compare how much easier it is to gain weight by eating one doughnut, than it is to shed calories by walking one mile. It is more time efficient and effective, to avoid eating the doughnut!

Those who are more seriously overweight, with Class II obesity (BMI between 30 and 40), or Class III, extreme obesity (BMI greater than 40), have a greater risk of death than average (three times more). For those with a BMI over 45, the risk of death is five times more on average, than it is for those of ideal weight. Their need to lose weight should be regarded life-saving, especially if they have coronary heart disease, high blood pressure and diabetes.

After attending to their 'Fat Mentality', they will need to reduce their portion sizes, reduce their calorie intake, limit their carbohydrate intake, and take care to choose foods that are good for their heart and circulation. The more disciplined they are, the easier it will be to accomplish weight loss. They should only add taxing exercise to their regime, after losing some weight. If these methods fail, they should be assessed medically for the most effective interventions.

This could be to prescribe antidepressants, appetite suppressants, semaglutide, CBT, or to undergo surgery.

Some with a highly developed 'Fat Mentality' mind-set, will benefit from regular mirror mantra exercises.

No-one can play the piano like Elton John, or the violin like Nigel Kennedy, without some training and frequent practice. They didn't awake one morning to find they could play their chosen instrument. Just as unlikely would be an overweight person who wakes one morning to find their weight was a bad dream. To accomplish weight loss, action not dreaming, is needed.

All musicians must be disciplined in their approach to practice, if their talent is to blossom. To lose weight and maintain the loss, the same principle must apply to eating habits. Those with a will of iron will have no need for this, or any other book to guide them. They can simply follow the advice to 'eat less, and exercise more!' They will know what to do. Others will need all the help they can get to defeat their 'Fat Mentality', their unnecessary eating habits and the self-gratification they desire.

Once those who are unhappily overweight have grasped all that can be done, and have put some of the methods I have detailed into practice, they will stand a better chance of achieving their weight loss goal.

Epilogue

Whether you are:

> Overweight and unhappy.
> Overweight and happy.
> Thin and sad, or
> Thin, fit and confident,

there is something to learn from understanding the 'Fat Mentality', and the 'Fit Mentality'. The question is, which mentality will you allow to dominate your eating, exercise habits and general health? The choice is yours.

Your quality of life is under your control; your longevity less so. Do not be persuaded that you will be happier, if you lose weight. It could be true, but this is a common psychological cop-up, used by those who think we should all be slim. Those who employ it will use their imagery, their preferences, and their preferred lifestyle to influence you.

Some will use population statistics to persuade you to lose weight (the government has been told they would save billions on medical expenses, if most overweight people lost weight). To counter this, many research results support the fact that obesity reduces longevity, so improving longevity will cost governments more in extra state pension payments.

Should we believe the statistics that predict a shorter life for obese individuals? To be specific, it is a group of overweight people that will not live as long (on average), as a group of slim people (all else being equal). Although this is likely to be true for each of us, no statistician will guarantee that a group statistic applies to any individual.

We can all benefit from asking ourselves what makes us happy, what will reduce our likeliness of illness, and what will prolong our life. If you have the answers, it is for you to decide what you are prepared to do to achieve them. This might mean ignoring the agendas of other people.

PART FIVE

APPENDICES

APPENDIX 1

WHAT SHOULD YOU WEIGH?

No-one can dictate what your weight 'should' be. It's your choice, but hopefully, after reading this book your choices will be better informed.

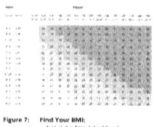

Figure 7: Find Your BMI: A Height / Weight Matrix

Population statistics and mortality rates clearly show health benefits for groups of ideal weight individuals. For groups, mortality and the occurrence of disease, increases with BMI. The greater the average BMI of a group of people, the greater their risk they have of diabetes and premature death.

Use your height (in feet, inches or meters), and cross reference it with your weight (in pounds or kilos) to determine your BMI (weight/height ratio2).

In the above diagram, the underweight are shown with a white background. They too, carry a greater risk of mortality than those of ideal weight (those BMIs with the lightest grey background).

Those in a slightly darker grey are overweight; the obese BMIs are shown with the darkest grey background. Anyone with a BMI greater than 40, is morbidly obese. That means that a group of such people can have a risk of dying prematurely, and a risk of getting diabetes and heart disease that is three to five times greater than a group of ideal weight people.

APPENDIX 2

OVERWEIGHT AND OUT OF BREATH? IS YOUR LIFE AT RISK?

There are heart symptoms to be taken seriously. If you would like to know more, read my book: 'How to Become Heart-Smart.' (2024).

Symptoms often occur late in the progress of heart and artery disease, but from the patient's point of view, trouble may seem to 'come out of the blue'. By occurring 'late', I mean that the disease process has been present and progressing for 10 – 20 years, before any symptom develops. We have an opportunity to detect disease early, and to do something about it before symptoms arise.

First Symptom or Last Symptom?

The consequences of heart and artery disease can be dramatic. Rarely, but significantly, our first symptom can be our last!

Symptoms to watch out for are:

- angina - chest tightness, pain or discomfort, occurring during exercise (occasionally occurring at rest);
- shortness of breath that is getting progressively worse,
- palpitation,
- blackouts and
- swollen feet.

It may come as a surprise, but severe chest pain is not the commonest presentation of heart disease. Angina, shortness of breath and 'running out of steam' with effort, are more common.

Combine obesity with diabetes, asthma, heart disease, arthritis and smoking, and the risks multiply. Remember that being underweight also has its risks.

APPENDIX 3

100 Calorie Food Portions & Carbohydrate Content

WEIGHT OF FOOD CONTAINING 100 CALORIES & CARBOHYDRATE CONTENT (Carbs)

FOOD GROUPS　　　　　　　　Grams of Food / Grams of Carbohydrate

ALCOHOL

BEER (average)	345	0.7
WINE (RED)	147	0.0
WINE:WHITE	152	0.9

BEVERAGES

TEA	unlimited	0.0
COFFEE	unlimited	0.0
COLA	2500	20.0
COLA (DIET)	unlimited	0.0
ORANGE JUICE	221	23.0
WATER	unlimited	0.0

BREAKFAST CEREAL

BRAN FLAKES	37	25.3
OATMEAL	27	17.5
RICE KRISPIES	36	32.9
SHREDDED WHEAT	26	19.0
MUESLI	28	12.7
CORNFLAKES	27	23.2
WHEETABIX	28	21.0
HONEY	30	10.6

DAIRY PRODUCE

COTTAGE CHEESE	99	0.0
EGG (WHITE)	278	0.0
EGG (YOLK)	29	0.0

MILK (1% FAT)	217	0.0
YOGURT	179	10.4

FISH

SALMON (PINK/BAKED)	47	0.0
HADDOCK	112	0.0
TUNA	100	0.0
HALIBUT	83	0.0
SOLE	110	0.0
COD	105	0.0
HERRING	55	0.0
MACKEREL	42	0.0
SARDINES	45	0.0
SWORDFISH	72	0.0
EEL	93	0.0

FRUIT

AVOCADO	53	0.2
STRAWBERRY	370	10.7
BANANA	86	15.7
APPLE	213	11.9
MANGO	148	26.1
PAPAYA	255	25.0

MEAT

TURKEY (BREAST ROASTED)   65   0.0
CHICKEN (DARK MEAT)   54   0.0
CHICKEN BREAST (ROASTED)   65   0.0
TURKEY (DARK MEAT)   56   0.0
DUCK (ROASTED)   51   0.0
PORK (LEG ROASTED)   55   0.0
BEEF (SIRLOIN) COOKED   56   0.0

NUTS

WALNUTS   15   0.5
SESAME SEEDS   17   0.1
HAZELNUTS   15   0.9
ALMONDS   16   1.1
SUNFLOWER SEEDS   18   0.6
SUNFLOWER FLOUR   17   3.1
PEANUT BUTTER   17   2.2
PEANUTS   18   2.2
TOFU   137   0.7
BRAZIL NUTS   15   0.5
PINE NUTS   15   0.6

OILS

OIL (WALNUT) 11 0.0
OIL (SUNFLOWER) 11 0.0
OIL (WHEATGERM) 11 0.0
OIL (CORN) 11 0.0
MAGARINE(POLYUN) 13 0.0
FLORA PRO-ACTIVE 7 0.0
OIL (HAZELNUT) 11 0.0
OIL (COD LIVER) 11 0.0
OIL (OLIVE) 11 0.0

OFFAL

LIVER (PIG) 53 1.9
LIVER (BEEF) 59 0.0
KIDNEY (BEEF) 72 0.0
LIVER (CHICKEN) 64 0.5
HEART 44 0.0

PULSES

LENTILS 95 15.5
SOYBEANS (DRY-ROASTED) 16 2.6
BLACKEYED PEAS 86 43.3

LIMA BEANS	30	18.8
BEANS (BAKED)	119	16.9

PREPARED FOODS

PASTA	35	19.0
SAUSAGES (PORK)	43	4.3
BREAD (WHOLEWHEAT)	46	18.2
BREAD (WHITE AVERAGE)	46	19.5

SHELLFISH

WHELK	111	0.0
OYSTERS	73	5.7
MUSSEL	96	0.0
CUTTLE FISH	100	0.0
CRAB (STEAMED)	78	0.0
SHRIMP	106	0.0
OCTOPUS	143	0.0
SCOLLOP	100	0.0
LOBSTER	97	0.0

VEGETABLES

POTATO (FLESH)	130	23.2
SPINACH	526	5.8
BRUSSEL SPROUTS	286	6.3
PEAS (GARDEN)	119	18.7
POTATO (FLESH & SKIN)	74	23.3
RICE (BROWN)	28	22.8
CORN	90	16.8
ASPARAGUS	400	2.8
PEPPER (GREEN)	190	18.5
LETTUCE	769	6.9
TOMATO	588	8.8
POTATO (SWEET)	115	13.3
BROCCOLI	183	1.7
CAULIFLOWER	357	4.3
ONION	278	11.1
CABBAGE	625	18.1
CARROTS	417	10.8
RICE (LONG-GRAIN WHITE)	28	22.2
PEAS (PROCESSED)	101	20.5

APPENDIX 4

Foods that Might Promote Artery 'Furring'

Grams of Food with 100KCal   /   Grams of Carbohydrate

CONFECTION

ICE CREAM	56	7.6
CHOCOLATE (MILK)	19	9.0
CHOCOLATE (PLAIN)	20	12.4
CHOCOLATE (WHITE)	17	18.3

DAIRY PRODUCE

PARMESAN	22	0.7

BUTTER	13	0.0
CHEDDAR	24	0.0
CAMEMBERT	34	0.0

FISH

| ANCHOVY | 52 | 0.0 |

MEAT

| LAMB (LEG ROASTED) | 45 | 0.0 |

FRUIT

| COCONUT | 17 | 0.9 |

OILS

| OIL (ALMOND) | 11 | 0.0 |
| OIL (PALM) | 11 | 0.0 |

PREPARED FOODS

| CRISPS (POTATO) | 22 | 11.6 |
| HAMBURGER | 41 | 8.7 |

CHEESEBURGER (triple meat)   39   8.2

Appendix 5

Foods Likely to be Good for Arteries and the Heart

CARDIAC VALUE OF FOODS (CV16 >0)

(Based on 16 Cardioprotective Nutrients)

Food	Portion Wt (g)	CV16
ALMONDS	28	7.4
ANCHOVY (in oil)	85	0.8
APPLE	138	1.5
ASPARAGUS	60	2.1
AVOCADO	100	3.4
BANANA	75	1.5
BEANS (BAKED)	100	2.2

BEEF (RUMP) COOKED	100	0.9
BEER (average)	500	1.9
BEVERAGES:COFFEE	200	1.3
BLACKEYED PEAS	80	5.2
BRAN FLAKES	50	3.1
BRAZIL NUTS	28	2.0
BREAD (WHITE AVERAGE)	30	1.1
BREAD (WHOLEWHEAT)	28	1.2
BROCCOLI	44	1.6
BRUSSEL SPROUTS	78	2.6
BUTTER	5	0.6
CABBAGE	75	1.5
CAMEMBERT	62	0.3
CARROTS	50	1.2
CAULIFLOWER	50	1.5
CHEDDAR	28	0.4
CHEESEBURGER (triple meat)	400	0.1
CHICKEN (DARK MEAT)	114	1.9
CHICKEN BREAST (ROASTED)	196	1.8
CHOCOLATE: MILK	44	0.2
COCONUT	23	0.2
COD	180	4.4
COLA	250	1.0
CORN	100	2.1
CORNFLAKES	100	1.2
COTTAGE CHEESE	100	2.0

CRAB (STEAMED)	118	8.3
CRISPS (POTATO)	30	0.4
CUTTLE FISH	85	9.7
DIET COLA	250	1.0
DUCK (ROASTED)	85	1.2
EEL	136	1.8
EGG (WHITE)	33	1.9
EGG (YOLK)	17	1.9
FLORA PRO-ACTIVE	5	1.5
HADDOCK	150	5.0
HALIBUT	150	4.7
HAMBURGER	250	0.3
HAZELNUTS	28	8.9
HEART (BEEF)	85	3.7
HERRING	143	3.6
HONEY	33	1.0
ICE CREAM	66	0.3
KIDNEY (BEEF)	85	16.4
LAMB (LEG ROASTED)	85	0.8
LENTILS	100	7.2
LETTUCE	55	2.0
LIMA BEANS	78	3.1
LIVER (BEEF)	85	28.5
LIVER (CHICKEN)	85	7.4
LIVER (PIG)	54	30.9
LOBSTER	145	2.8

MACKEREL	88	3.5
MAGARINE	5	1.6
MANGO	100	1.4
MILK (1% FAT)	245	1.4
MILKSHAKE	200	2.5
MUSSEL	85	11.7
OATMEAL	234	2.5
OCTOPUS	85	4.9
OIL (ALMOND)	14	0.7
OIL (COD LIVER)	14	0.9
OIL (CORN)	14	4.3
OIL (HAZELNUT)	14	1.4
OIL (OLIVE)	14	0.8
OIL (PALM)	14	0.7
OIL (SUNFLOWER)	14	6.0
OIL (WALNUT)	14	6.3
OIL (WHEATGERM)	14	4.3
ONION	50	1.5
ORANGE JUICE	124	1.3
OVALTINE	200	2.4
OYSTERS	70	12.4
PAPAYA	100	1.3
PARMESAN	6	0.7
PASTA	57	1.9
PEANUT BUTTER	32	3.5
PEANUTS	28	3.3

PEAS (GARDEN)	80	2.5
PEAS (PROCESSED)	80	1.2
PEPPER (GREEN)	74	2.0
PINE NUTS	28	1.4
PLAIN CHOCOLATE	44	0.1
PORK (LEG ROASTED)	85	0.9
POTATO (FLESH & SKIN)	200	2.4
POTATO (FLESH)	100	3.5
POTATO (SWEET)	100	1.6
RED WINE	100	1.8
RICE (BROWN)	100	2.1
RICE (LONG-GRAIN WHITE)	100	1.2
RICE KRISPIES	25	1.8
SALMON (PINK/BAKED)	150	6.1
SARDINES (in oil)	50	3.5
SAUSAGES (PORK)	100	1.4
SCOLLOP	120	4.8
SESAME SEEDS	28	9.8
SHREDDED WHEAT	50	1.7
SHRIMP	50	5.4
SOLE	127	4.4
SOYBEANS (DRY-ROASTED)	50	7.1
SPINACH	100	2.9
STRAWBERRY	83	1.7
SUNFLOWER FLOUR	28	3.5
SUNFLOWER SEEDS	28	6.7

SWORDFISH	106	2.9
TEA	200	1.5
TOFU	100	2.4
TOMATO	100	1.7
TUNA	150	4.8
TURKEY (BREAST ROASTED)	85	2.0
TURKEY (DARK MEAT)	85	1.6
WALNUTS	28	21.2
WATER	**1000**	**1.0**
WHELK	85	13.9
WHITE CHOCOLATE	44	0.1
WHITE WINE	100	1.6
YOGURT	100	1.1

APPENDIX 6

SALT (SODIUM) UNITS IN FOODS

FOOD   PORTION WEIGHT (g)   Salt Units
(One Daily Requirement = 100 Units)

VEGETABLES

ASPARAGUS   60   0
BROCCOLI   44   0
BRUSSEL SPROUTS   78   0
CABBAGE   75   0
CARROTS   50   2
CAULIFLOWER   50   0
CORN   100   0
LETTUCE   55   0
ONION   50   0
PEAS (GARDEN)   80   0

PEAS (PROCESSED)   80   19
PEPPER (GREEN)   74   0
POTATO (FLESH & SKIN)   200   2
POTATO (FLESH)   100   0
POTATO (SWEET)   100   2
RICE (BROWN)   100   0
RICE (LONG-GRAIN WHITE)   100   0
SPINACH   100   8
TOMATO   100   1

SEAFOOD

CRAB (STEAMED)   118   31
CUTTLE FISH   85   5
LOBSTER   145   30
MUSSEL   85   19
OCTOPUS   85   10
OYSTERS   70   0
PRAWN   50   50
SCOLLOP   120   0
SHRIMP   50   12
WHELK   85   15

BREAD, PASTA, CRISPS

BREAD (WHITE AVERAGE)   30   9

BREAD (WHOLEWHEAT)	28	10
CRISPS (POTATO)	30	14
PASTA	57	1

BEANS

BEANS (BAKED)	100	33
BLACKEYED PEAS	80	1
LENTILS	100	0
LIMA BEANS	78	20
SOYBEANS (DRY-ROASTED)	50	0

OFFAL

HEART	85	4
KIDNEY (BEEF)	85	8
LIVER (BEEF)	85	8
LIVER (CHICKEN)	85	4
LIVER (PIG)	54	4
PANCREAS	85	0

OILS

FLORA PRO-ACTIVE	5	0
MAGARINE (POLYUN)	5	3
OIL (ALMOND)	14	0

OIL (COD LIVER)   14   0
OIL (CORN)   14   0
OIL (HAZELNUT)   14   0
OIL (OLIVE)   14   0
OIL (PALM)   14   0
OIL (SUNFLOWER)   14   0
OIL (WALNUT)   14   0
OIL (WHEATGERM)   14   0

NUTS AND SEEDS

ALMONDS   28   0
BRAZIL NUTS   28   0
COCONUT   23   0
HAZELNUTS   28   0
PEANUT BUTTER   32   7
PEANUTS   28   0
PINE NUTS   28   0
SESAME SEEDS   28   0
SUNFLOWER FLOUR   28   0
SUNFLOWER SEEDS   28   0
TOFU   100   0
WALNUTS   28   0

MEAT PRODUCTS

BEEF (SIRLOIN) COOKED   100   5
CHEESEBURGER (triple meat)   400   170
CHICKEN (DARK MEAT)   114   4
CHICKEN BREAST (ROASTED)   196   10
DUCK (ROASTED)   85   5
HAMBURGER   250   97
LAMB (LEG ROASTED)   85   3
PORK (LEG ROASTED)   85   4
SAUSAGES (PORK)   100   75
TURKEY (BREAST ROASTED)   85   3
TURKEY (DARK MEAT)   85   6

FRUITS

APPLE   138   0
AVOCADO   100   0
BANANA   75   0
MANGO   100   0
ORANGE JUICE   124   1
PAPAYA   100   0
STRAWBERRY   83   0

FISH

ANCHOVY 85 209
COD 180 18
EEL 140 58
HADDOCK 150 7
HALIBUT 150 7
HERRING 143 14
MACKEREL 88 3
SALMON (PINK/BAKED) 150 5
SARDINES 50 14
SOLE 127 10
SWORDFISH 106 11
TUNA 150 30

DAIRY PRODUCE

BUTTER 5 2
CAMEMBERT 62 23
CHEDDAR 28 13
COTTAGE CHEESE 100 19
EGG (WHITE) 33 4
EGG (YOLK) 17 1
MILK (1% FAT) 245 7
PARMESAN 6 3
YOGURT 100 5

CONFECTION

CHOCOLATE:(PLAIN)   44   0
CHOCOLATE:(WHITE)   44   3
CHOCOLATE(MILK)   44   2
ICE CREAM   66   2

BREAKFAST CEREAL

BRAN FLAKES   50   25
CORNFLAKES   100   63
HONEY   33   0
MUESLI   100   3
OATMEAL   240   85
RICE KRISPIES   25   10
SHREDDED WHEAT   50   0
WEETABIX   50   8

ALCOHOL

BEER (average)   500   2
WINE: RED   100   0
WINE:WHITE   100   0

BEVERIDGES

COFFEE 200 0
DIET COLA 250 0
MILKSHAKE (Vanilla) 200 0
OVALTINE 200 0
TEA 200 0
WATER 1000 0

BIBLIOGRAPHY

Aoun, A., Darwish, F., Hamod, N. The Influence of the Gut Microbiome on Obesity in Adults and the Role of Probiotics, Prebiotics, and Synbiotics for Weight Loss. Prev. Nutri. Food Sci. 2020 Jun 30; 25(2): 113–123. 30. doi: 10.3746/pnf.2020.25.2.113

Bikman, B.T., Fisher-Wellman, K.H.(2021). The Metabolic Effects of Ketones. Int. J. Mol. Sci. 22(15), 8292.

Brogan, Caroline (31. 1. 2017). Early death and ill-health linked to low socioeconomic status. Imperial (College, London) News.

BMI and all-cause mortality: systematic review and non-linear dose-response meta-analysis of 230 cohort studies with 3.74 million deaths among 30.3 million participants. British Medical Journal: 2016; 353: i2156. doi: 10.1136/bmj.i2156).

Carmienke, S. et al. 2013). General and abdominal obesity parameters and their combination in relation to mortality: a systematic review and meta-regression analysis. Eur. J. Clin. Nutr. 67, 573–585.

Chaouachi, M. et al. (2024). A Review of the Health-Promoting Properties of Spirulina with a focus on Athletes' Performance and Recovery. J Diet Supplement. 21(2):210-241. doi:10.1080/19390211.2023.2208663. Epub 2023 May 4.

Chen, I-Ju et al. (2016) Therapeutic effect of high-dose green tea extract on weight reduction: A randomised, double-blind, placebo-controlled clinical trial. Clin Nutrition. Jun; 35(3): 592-9.
doi: 10.1016/j.clnu.2015.05.003. Epub 2015 May 29.

Clemens Drenowatz, Gregory A. Hand, Robin P. Shook, John M. Jakicic, James R. Hebert, Stephanie Burgess, and Steven N. Blair. (2015).The association between different types of exercise and energy expenditure in young non-overweight and overweight adults. Applied Physiol. Nutrition and Metabolism. 40(3); 211-217. Published online 2014 Oct 29. doi: 10.1139/apnm-2014-0310

Deemer, S.E., Plaisance, E.P., Martins, C. (2020). Impact of Ketosis on Appetite Regulation – a Review. https://doi.org/10.1016/j.nutres.2020.02.010Get rights and content.

De Logeril, M., Salen, P., Martin, J.L. et al. The Lyon Diet. Mediterranean diet, traditional risk factors, and the rate of cardiovascular complications after myocardial infarction: final report of the Lyon Diet Heart Study. Circulation (1999): 99(6); 779-785.

European Medicines Agency. Wegovy®: Reference Number: EMA/162766/2023.

Ford, ES, Caspersen, CJ. Sedentary behaviour and cardiovascular disease: a review of prospective studies. Int J Epidemiol. 2012; 41:1338–1353. doi: 10.1093/ije/dys078

Harvard Health Publishing. Harvard Medical School. Calories burned in 30 minutes for people of three different weights. March 8, 2021.

Hong, Y. et al. 2007. Metabolic syndrome, its preeminent clusters, incident coronary heart disease and all-cause mortality: results of prospective analysis for the atherosclerosis risk in communities study. J. Intern. Med. 262, 113–122.

Jain, S., Singh, S.N. (2013) Metabolic effect of short term administration of Hoodia gordonii, an herbal appetite suppressant. South African Journal of Botany. Volume 86, 51-55. https://doi.org/10.1016/j.sajb.2013.02.002

Kahnman, Daniel (2011). *Thinking Fast and Slow*. Macmillan.

Lichtman SW, Pisarska K, Berman ER, Pestone M, Dowling H, Offenbacher E, Weisel H, Heshka S, Matthews DE, Heymsfield SB: *Discrepancy between self-reported and actual caloric intake and exercise in obese subjects.* N Engl J Med. 1992; 327: 1893– 1898.

National Heart, Lung and Blood Institute. *Morbidity & Mortality (1998) chartbook on cardiovascular, lung, and blood diseases.* Rockville, Maryland: US Department of Health and Human Services, National Institutes of Health.

Rosenman, R.H., Friedman. M. (1974). *Type A Behaviour and Your Heart.* Knopf Doubleday Publishing Group.

Shook, R.P., Blair, S.N., Duperly, J. et al (2014) *What is Causing the Worldwide Rise in Body Weight?*
European Endocrinology. Aug; 10(2): 136-144.
Cited in US National Health and Nutrition Examination Surveys (2011).

Samitz G, Egger M, Zwahlen M. *Domains of physical activity and all-cause mortality: systematic review and dose-response meta-analysis of cohort studies.* Int J Epidemiol. 2011;40:1382–1400. doi: 10.1093/ije/dyr112.

Stunkard, A.J., Harris, J.R., Pedersen, N.L. et al (1990) *The Body-Mass Index of Twins Who Have Been Reared Apart.* New England J Med; 322:1483-1487.

Suyog, S. Jain, Sunita J. Ramanand, Jaiprakash B. Ramanand, Pramod B. Akat, Milind H. Patwardhan, and Sachin R. Joshi. (2011) *Evaluation of efficacy and safety of orlistat in obese patients.* Indian J. Endocrinology and Metabolism. Apr-Jun; 15(2): 99–104. doi: 10.4103/2230-8210.81938

Wansink, Brian (2006). *Mindless Eating: Why We Eat More than We Think.* Bantam.

Yuan, Ma, Feng, J. He, MacGregor. G.A. (2015). *High salt intake: independent risk factor for obesity?*
Hypertension. 66(4):843-9. doi:10.1161/HYPERTENSIONAHA.115.05948. Epub 2015 Aug 3.

Ziółkiewicz, A., Kasprzak-Drozd, K. et al. (2023). *The Influence of Polyphenols on Atherosclerosis Development.* Int J Mol Sci. 24(8): 7146

INDEX

1
100 calorie portions, 151, 152, 153
5
5HT, 240
A
abdomen, 6, 58, 65, 98
abdominal fat, 26, 56
abdominal girth, 56
Acadamenes, 13
academic, 13
Acceptance and Commitment Therapy, 242
accumulating fund, 40
acromegaly, 98
active, 3, 25, 29, 107, 151, 238
activity, 2, 38, 107, 133, 156, 157, 232

addictive substance, 37, 156

adversity, 27

advertising, 22, 83, 119

aerobic ability, 207

aerobic fitness, 15

afternoon tea, 194, 197

AI-based, 140

alcohol, 25, 37, 111, 112, 128, 134, 137, 199, 200, 222, 241

all-inclusive, 39, 183

amino-acids, 168

Aminorex, 236

Amish, 11

amphetamine, 235, 236

Amy Winehouse, 45

anaemia, 62

anathema, 19

ancestry, 9

anchovy, 227

anecdotal information, 71

angina, 6, 64, 167, 169, 170, 255

animal experiments, 168

Anne Widdicome, 79

anorexic, 60, 147, 195, 198, 199

antidepressant medications, 92, 230

anxious, 194, 198

aphorisms, 15

Apollo, 10, 30
appearance, 31, 32, 83, 98, 111, 163, 199
appetite, 1, 2, 19, 22, 87, 91, 92, 93, 95, 96, 114, 118, 119, 121, 122, 125, 128, 129, 132, 135, 140, 146, 151, 158, 164, 165, 170, 179, 187, 192, 196, 217, 224, 230, 231, 232, 233, 235, 236, 237, 240, 247, 282
appetite Cycle, 129
arginine, 171
arm twisting, 5, 47, 110
art, 55, 88, 197, 216
artery health, 6, 169, 170
artificial sweeteners, 158, 225
asthma, 89, 98, 255
Athenians, 10, 84
atheroma, 168, 171
atherosclerosis, 64, 167, 168, 169, 171, 282
athletic, 2, 17, 166, 221
Atkin's, 43, 48, 160
attitudes, 14, 40, 44, 184, 241, 242
attractive, 45, 57, 83, 112, 134
average, 15, 52, 55, 56, 60, 70, 74, 76, 77, 78, 87, 89, 110, 161, 180, 195, 203, 204, 240, 246, 249, 252, 256, 267, 278
avoid failure, 106
avoid seeing yourself, 50

B

Bacchus, 10
bacteria, 232, 233, 234

bacterium, 231
baked beans, 227
ballroom dancing, 203
battles, 22, 24
beauty, 18, 176
Beers, 222
behaviour patterns, 40, 184
behavioural characteristics, 93
benzphetamine, 236
best we can do, 71, 110
better educated, 7, 78
better off, 7
Bifidobacterium longum, 233
Big Dipper' test, 16
biological, 6, 30, 37, 45, 82, 94, 166, 187
biology, 56, 94, 187
biscuits, 122, 144, 159, 223
blackouts, 255
blame, 27, 187
blood pressure, 72, 74, 89, 109, 110, 157, 171, 172, 193, 225, 226, 227, 234, 235, 238, 246
blood sugar, 158, 223
BMI, (Body Mass Index) 54, 55, 57, 74, 75, 76, 77, 78, 203, 204, 229, 237, 238, 239, 240, 242, 243, 244, 245, 246, 250, 252, 280
Bodhi, 126
bodily characteristics, 55

body and mind, 183
body fat, 26, 57, 63, 74, 96, 156, 164, 184, 217, 220
body shape, 30, 60, 61, 65, 92
body weight, 2, 7, 9, 11, 27, 30, 54, 55, 56, 63, 64, 65, 75, 77, 89, 91, 93, 95, 145, 157, 181, 197, 198, 200, 203, 217, 231, 232, 238
bodyweight, 15, 56, 70, 71, 74, 75, 76, 77, 78, 92, 203, 228
bodyweight and shape, 56
bodyweight of adults, 70
bones visible, 66
boot camp, 18, 19
bored, 14, 126, 127, 185, 197, 207
boredom, 125, 205
bowel bacteria, 231
bowel function, 158, 232
bowel inflammation, 232
brain, 6, 19, 50, 86, 91, 92, 93, 95, 128, 133, 165, 232, 236, 240
bread,, 144, 156, 159
breakfast, 137, 165, 190, 191, 203, 219, 224, 225
breathing, 62, 244
breathless, 5, 47, 61, 62, 63, 64, 66, 67, 112, 205, 206, 207
Bridget Jones, 111
British population, 70
brown fat, 81, 82
Buddha, 126
bulimia, 59, 198
bullying, 5, 45, 47, 69

buns, 144

bureaucrats, 134, 237

Bureauniks, 13

burn, 12, 81, 107, 121, 156, 157, 158, 159, 181, 191, 203, 217, 220, 222, 240

burning of body fat, 164

business, 11, 25, 26, 40, 80, 159, 180, 184, 185

Butterfly Effect, 94, 180

buxom, 45

bypass operation, 167

C

cafes, 119

cafeterias, 119

cakes, 144, 145, 156, 159

calcium, 168, 169, 226, 234

CaloAI®, 140, 159

caloric intake, 283

calorie intake, 85, 94, 151, 246

calories, 2, 3, 4, 6, 44, 77, 81, 82, 107, 114, 121, 130, 135, 138, 143, 145, 146, 147, 148, 149, 151, 152, 153, 156, 157, 158, 162, 178, 181, 191, 203, 204, 217, 218, 220, 221, 222, 230, 239, 246

camaraderie, 50

cancer, 7, 48, 73, 81, 89, 165, 169, 170, 228

candidate, 15, 16

capitalism, 12

carbohydrate, 3, 7, 17, 19, 83, 115, 138, 141, 145, 151, 155, 156, 157, 158, 159, 160, 161, 162, 163, 164, 165, 173, 178, 181, 217, 218, 220, 222, 225, 230, 233, 240, 245, 246

carbohydrates, 1, 4, 114, 121, 156, 158, 162

carbs, 4, 44, 114, 156, 157, 158, 163, 165, 183, 195, 225

cardiovascular disease, 77, 282

cathartic, 20

CBT (Cognitive Behavioural Therapy), 241

Celebrity, 17

central nervous system, 240

challenge, 17, 27, 97, 166, 187, 188, 213

charisma, 18

charlatan, 50

chefs, 27, 163, 199

chemistry, 19

chest tightness, 255

chew, 197

children, 65, 70, 98, 107, 123, 192

chocolate, 19, 122, 145, 159

choice, 3, 16, 29, 44, 83, 84, 90, 119, 138, 176, 179, 186, 222, 248, 251

cholesterol, 64, 89, 110, 169, 170, 172

chronic disease, 56, 202

chubby, 45

circuits, 205

circulation, 48, 73, 115, 161, 172, 173, 246

citizen, 13, 15, 17, 32
citizenship, 15, 16, 19
class-defining, 194
clinical organisations, 88
clot (blood), 169
clothes, 5, 12, 44, 46, 50, 54, 56, 61, 97, 111
clothes' size, 54, 55, 56
cognitive therapy, 38
collaborate, 44
colon, 128, 170, 232
comfort and luxury, 39
comfortable, 5, 21, 43, 47, 90, 118, 130, 136
commercial, 14, 119
commitment, 49, 95, 177, 221, 242, 245
companions, 50, 188
compatibility, 20
compliance, 193
compulsion, 26, 144, 187, 188
confectionary, 163
conquering their fear, 123
constipation, 118, 128, 238
contraceptive pill, 96, 97
control, 1, 2, 3, 4, 6, 23, 24, 26, 27, 29, 30, 36, 37, 40, 47, 84, 91, 93, 95, 96, 97, 98, 105, 114, 118, 128, 137, 138, 179, 182, 190, 193, 223, 230, 232, 233, 241, 243, 245, 248
controlling, 6, 23, 26, 145, 175, 200

controversial subjects, 14
convenience food, 145
cope with life, 51
cop-out, 100, 192
coronary arteries, 167
coronary artery disease, 56, 64, 73, 89, 169, 171
counselling, 38, 86, 87, 146, 230
courage, 50, 185
creases (skin), 57, 58, 61, 65, 66, 67
cultural conflict, 20
cultural influences, 44, 84, 93
culture, 5, 6, 9, 17, 18, 20, 81, 84, 194
curvy, 45
Cushing's Syndrome, 98
customers, 11, 191
cycling, 204

D

danger., 72
death, 14, 55, 59, 76, 77, 89, 124, 169, 237, 246, 252, 280
defeat, 23, 100, 101, 137, 139, 247
dehydration, 128, 208
deleterious effects, 45, 171
demeanour, 10, 214
denial, 32, 51, 69
depersonalised group, 55
depression, 51, 106, 112, 124, 161, 230, 236

deprivation, 19, 37, 123, 197, 198

derogatory, 11

desire, 2, 5, 16, 19, 24, 46, 88, 92, 101, 118, 119, 122, 129, 137, 156, 181, 187, 209, 243, 247

dexfluramine, 236

diabetes (Type 2), 4, 46, 72, 74, 90, 164, 228, 229, 238, 239, 244, 246, 252, 253, 255

diet regime, 27

diethylpropion, 235

dieting, 2, 4, 21, 24, 48, 52, 61, 79, 85, 86, 95, 102, 103, 104, 106, 116, 123, 133, 143, 155, 160, 209, 213, 240, 242

diets, 21, 119, 165, 218

Dionysus, 10

disadvantaged, 7, 158

discipline, 10, 14, 24, 49, 87, 106, 112, 143, 151, 209, 213, 214, 223, 245

disciplined, 2, 17, 20, 143, 185, 209, 213, 214, 246, 247

discover, 73, 127

disfigurement, 31

disharmony, 10

disheartened, 32

disinterested, 4, 32

disrespectful, 57

distance walking, 204

distorted perception, 54. 60, 179

diuretics, 227, 234, 235

DNP, 240

doctors, 52, 72, 98, 110, 227, 234

dominant, 22, 29, 36, 37, 178, 179

dominant mentality, 29, 37

dopamine, 92

Dorothy (Wizard of Oz), 49, 50

doughnuts, 30, 34, 80, 145, 181, 182

Dr. Atkins' Diet, 120

dramas, 29

dreaming, 191, 221, 247

dress code, 193, 214

dress sizes, 56

dressing, 11

drugs, 48, 64, 111, 112, 128, 234, 237, 238

Duke of Wellington, 23

duodenal switch operation, 244

duty, 11, 214

E

eat sparingly, 94, 190

eating, 1, 2, 3, 4, 5, 17, 18, 19, 21, 22, 23, 24, 27, 30, 36, 37, 38, 39, 69, 79, 80, 81, 82, 83, 85, 93, 96, 101, 105, 114, 119, 121, 122, 124, 127, 128, 134, 136, 138, 139, 140, 141, 143, 151, 157, 159, 160, 162, 164, 166, 170, 171, 182, 183, 189, 190, 191, 192, 193, 194, 195, 196, 197, 198, 217, 219, 221, 223, 224, 246, 247, 248

Eating and conversation, 197

Eating Disorders, 195

Eating Habits and Culture, 189
economy, 183
educated, 78, 229
education, 7, 13, 14, 74, 78
Edward Lorenz, 94
ego, 113, 185
elephant in a room, 2
embarrassment, 32
emergency admissions, 72
emotion, 29, 169
emotional blackmail, 109
endocrinologist, 97
energy, 6, 10, 18, 25, 26, 30, 39, 44, 58, 59, 81, 84, 86, 93, 95, 106, 107, 121, 124, 133, 137, 138, 147, 148, 156, 157, 158, 165, 166, 181, 182, 188, 191, 193, 197, 203, 207, 217, 220, 221, 233, 235, 281
environment, 37, 199
environmentally conscious, 199
equality, 12, 17, 45
ethnicity, 74
etiquette (eating), 193, 194
evolution, 12, 203
evolutionary advantage, 6
evolutionary demand, 45
exceptional, 56
excess, 6, 25, 26, 39, 57, 59, 63, 82, 96, 119, 156, 220, 240
excessive weight gain, 52, 89

excuses, 119, 188

exercise, 1, 2, 3, 4, 6, 11, 13, 14, 20, 21, 24, 27, 32, 36, 37, 38, 39, 44, 45, 47, 57, 63, 64, 67, 79, 80, 81, 83, 84, 86, 92, 99, 100, 105, 121, 128, 133, 137, 151, 152, 156, 159, 179, 181, 183, 191, 202, 203, 204, 205, 206, 207, 208, 209, 216, 217, 220, 221, 222, 223,230, 234, 235, 241, 242, 246, 247, 248, 255, 281, 283

exhausting, 126

expectations, 44, 55, 84, 88

experiments, 85, 168, 171, 216

explosive mixture, 240

extra chins, 66

extreme obesity, 75

F

fable, 9

failed to lose weight, 30, 80, 81, 86, 105, 180

faint, 138, 226, 234

fake news, 73

family, 6, 11, 12, 16, 25, 26, 27, 54, 62, 68, 69, 73, 84, 88, 170, 171, 173, 198, 236, 242

famine, 6, 7, 26, 58, 82, 85, 158, 196

Fanny and Johnny Craddock, 193

fantasy, 13, 21, 22

fashion, 25, 46, 59, 68, 76, 83

fashionable, 5, 46

fasting, 37, 124, 127, 130, 140, 181, 221

fat cat, 25, 26

Fat Mentality, 1, 2, 3, 4, 6, 9, 10, 11, 16, 19, 21, 22, 23, 24, 25, 26, 27, 30, 31, 32, 35, 36, 37, 38, 40, 47, 94, 99, 100, 101, 107, 114, 119, 120, 121, 123, 124, 125, 127, 137, 138, 141, 142, 143, 144, 159, 160, 166, 175, 177, 178, 179, 180, 182, 187, 188, 192, 223, 243, 245, 246, 247, 248

  fat reserves, 26

  fattening, 145, 200, 222

  fatty acids, 232, 233

  fatty meat, 77

  fear of hunger, 122

  feasting, 11

  feel healthier, 112

  feeling of fullness, 91, 93, 96, 237, 243

  female, 51, 96, 98, 184

  fibre, 158, 170, 224, 232

  fickle, 111

  fighting, 6, 13, 14, 24, 214

  financial, 7, 16, 25, 26, 134, 184

  finding a partner, 45

  first step, 23, 100, 114, 175

  fish, 17, 77, 152, 162, 172, 183, 191, 193

  fit, 2, 4, 12, 24, 25, 28, 30, 37, 49, 56, 61, 63, 64, 79, 83, 84, 93, 94, 100, 105, 111, 112, 175, 180, 182, 184, 185, 187, 188, 195, 196, 197, 204, 206, 208, 227, 242, 248

Fit Mentality, 1, 2, 3, 4, 9, 10, 13, 23, 24, 27, 30, 31, 35, 36, 37, 38, 40, 86, 94, 100, 107, 114, 121, 166, 175, 180, 181, 182, 183, 184, 185, 188, 197, 242, 248

fitness level, 44, 80, 220

Fitrakia, 9, 11, 12, 13, 15, 17, 18, 20, 21, 23, 41

Fitrakian, 12, 13, 14, 15, 18, 20, 166

fitter, 1, 84, 182, 206

flavonoids, 224

fluid retention, 96

food, 1, 3, 6, 7, 11, 12, 14, 16, 17, 18, 19, 20, 21, 23, 28, 29, 30, 32, 33, 36, 37, 38, 39, 44, 47, 48, 50, 53, 59, 73, 74, 80, 81, 82, 83, 84, 85, 86, 87, 91, 92, 93, 96, 99, 105, 115, 118, 119, 120, 121, 123, 124, 125, 126, 128, 129, 130, 132, 133, 134, 135, 136, 137, 138, 139, 140, 141, 142, 143, 144, 145, 146, 148, 149, 151, 152, 153, 156, 157, 158, 159, 160, 161, 162, 163, 164, 165, 166, 167, 168, 169, 170, 171, 172, 173, 174, 178, 179, 182, 183, 184, 187, 188, 189, 190, 192, 195, 196, 197, 198, 199, 200, 202, 203, 217, 218, 219, 220, 221, 223, 224, 225, 226, 227, 228, 230, 231, 232, 233, 234, 237, 241, 242, 243, 244, 245, 246

food choice, 29, 160

food companies, 119

food labelling, 173

food promoters, 119

foraging, 188, 232

Forest Gump, 37

formula, 48

friends, 27, 30, 50, 54, 68, 84, 113, 196, 198, 228

fructose, 158, 223, 224

fruit, 17, 77, 83, 145, 158, 166, 223, 224, 231

frying, 168

fuel, 17, 107, 148, 156, 166, 183, 207, 220

fulfillment, 124, 184

furring (artery), 6, 64, 73, 97, 120, 167, 168, 169, 171, 172, 224

G

game, 113, 185

Garcinia cambogia, 231

gastric banding, 243

gastric bypass, 243

gene, 92

general health, 115, 173, 246, 248

generation and attitude, 123

genes, 1, 3, 6, 10, 15, 19, 20, 37, 91, 92, 93, 94, 128, 169, 223

Genetic analysis, 163

genetic code, 19

genetic factor, 92

genetic influences, 92, 93

genetic makeup, 93

genetic predisposition, 156

genetic profile, 2, 4, 71

George Bernard Shaw, 118

ghrelin, 95, 165, 232

giant ogre, 100

GLP-1, 95, 96, 232, 237

glucose, 158, 238
gluten-free, 218
glutton, 30, 128, 147, 193, 199
gluttons, 195
gluttony, 45, 57, 128
glycaemic, 114, 158, 164, 223, 224
glycogen, 157, 158
go large, 39, 83
grace, 112, 194, 197
graceful aging, 111
grandmother's genes, 71
grapes, 224
grass, 163
gratification, 1, 9, 10, 14, 19, 37, 49, 81, 128, 137, 144, 178, 183, 188, 190, 247
grazing, 39, 195, 197
greater risk of dying, 57, 70
greed, 26, 125, 187, 188
greedy, 100, 189
green tea, 118, 224, 232, 281
gross (morbid) obesity, 73
grossly overweight, 65
group therapy, 230
groups of people, 57
growth hormone, 96, 98
guarantee, 46, 89, 110, 249

guilt, 36, 145

gurus, 50

gut biome, 3, 6, 231, 233

Gut Biome, 232

gut hormones, 1, 3, 6, 81, 91, 93, 223, 232, 233

gym, 31, 182, 184, 189, 207, 217

gymnasium, 4, 12, 15, 206

H

habit, 37, 122, 157, 199, 235

habits, 5, 21, 22, 23, 24, 27, 49, 69, 84, 105, 145, 188, 190, 191, 200, 247, 248

hair growth, 98

half portions, 152

happiness, 9, 10, 20, 166, 184, 185, 193

happy-go-lucky, 10, 17

hardship, 37, 49

HDL cholesterol, 64

head in the sand, 51

health benefits, 45, 46, 87, 217, 252

health divide, 7, 228

health issues, 15

health of our arteries, 3

health professionals, 83

health risk, 5, 47

health services, 7

heart, 2, 3, 4, 6, 7, 25, 28, 46, 48, 49, 50, 53, 59, 62, 63, 64, 72, 73, 74, 78, 86, 90, 97, 110, 115, 120, 159, 160, 161, 165, 166, 167, 169, 170, 171, 172, 173, 174, 193, 196, 228, 229, 230, 235, 238, 240, 244, 246, 253, 254, 255, 282

heart attacks, 6, 25, 46, 64, 72, 78, 110, 120, 167, 170, 172

heart disease, 4, 7, 28, 59, 62, 72, 73, 74, 90, 97, 120, 160, 170, 171, 173, 193, 228, 229, 244, 246, 253, 255, 282

heart muscle, 169

Heart-Smart, 62, 64, 171, 254

height and weight calculations, 55

heptral, 241

high blood pressure, 172, 226

high fibre, 96, 145

high-carb snacks, 144

high-intensity endurance training, 204

historic examples, 84

Holy Grail, 80

Home delivery, 12

homes, 10, 12

hoodia, 118, 232, 282

hormone, 43, 95, 96, 97, 98, 164, 165, 221, 232, 237

hormones, 3, 19, 44, 47, 86, 96, 97, 98, 128, 221, 226, 235, 241

hot flushes, 96

how we eat, 194

human, 12, 30, 56, 70, 110, 164, 187, 203, 233

humanity, 187, 228

hunger, 93, 96, 119, 122, 123, 124, 125, 127, 136, 144, 187, 209, 224, 232, 235, 237

hungry, 19, 22, 23, 27, 39, 80, 85, 118, 120, 122, 123, 124, 125, 126, 128, 136, 141, 142, 144, 146, 177, 182, 183, 188, 195

hunter gatherers, 202

hypertension, 171, 172, 226

hypnosis, 38

hypothalamus, 95, 240

I

ideal weight, 46, 54, 60, 69, 70, 72, 76, 89, 94, 204, 246, 252, 253

Ideal Weight, 67

ideal weight', 60

ignorance,, 51

image, 69, 176, 179

inactivity, 2, 176, 191

inclination, 39, 44

indigestion, 192

individual difference, 12

indoctrinated, 191

indolence, 45, 203

inducements on offer, 83

indulge, 31, 48, 143

inequality, 7

influence, 2, 37, 78, 83, 92, 93, 94, 96, 97, 121, 160, 223, 232, 233, 248

inheritance, 18, 25, 60, 65, 91

inherited trait, 92

injection, 22, 48, 80

insecurity, 185, 194

instinctive, 29

insulin, 96, 164, 231, 238, 239

intellectual, 14, 166

Internet, 72

intragastric balloon, 243

iodine, 97, 226, 234

IQ puzzles, 15

J

juicing, 218

K

Kahnman, 29, 283

ketone, 165, 217

ketones on your breath, 165

Ketostix®, 164

Kim's Case, 63

know-how, 113

knowledge, 9, 28, 71, 72, 85, 88, 105, 113, 226

L

Lactobacillus acidophilus, 233, 234

large arms, 66

large portions, 39, 183

large-scale group studies, 89

late-night snacks, 197

laziness, 23, 24, 57
learning, 10, 13, 137
legal procedure, 73
lessons, 112, 137, 144, 216
life expectancy, 48, 70
lifestyle, 3, 4, 17, 20, 22, 24, 28, 60, 84, 90, 93, 137, 223, 248
like-minded, 19, 20, 242
lipase inhibitor, 239
liraglutide, 239
long distance walking, 82
longevity, 73, 88, 89, 248, 249
long-term health, 87, 217
losing weight, 6, 23, 32, 61, 69, 70, 79, 85, 99, 100, 109, 111, 120, 121, 160, 176, 182, 213
losing weight, 21, 48, 86, 97, 99, 100
love affair, 155
low-carb diet, 43, 158, 160, 161, 165, 171, 217, 223, 224, 225
lungs, 2, 62, 64, 283
Lyon diet, 172

M

magic pill, 43
magnesium, 170, 226, 234
maintain the loss, 22, 47, 247
making money, 25
males, 58, 204
mañana, 106

manganese, 170, 226

manipulative relationship, 109

marathon, 156

marriage, 143

MC4R, 92, 239

meals, 38, 183, 195, 197, 198, 239

mealtimes, 17, 190

medals, 18, 100

media, 45, 72, 73, 137, 157, 238

medical advice, 88

medical appraisal, 122

medical expenses, 249

medical help, 2, 4, 230

medical practice, 88

medical professionals, 2, 5, 55, 46, 47, 72

medical research, 44

Mediterranean diet, 77, 282

menopause, 96, 97, 226, 235

menstrual disorders, 98

mental concentration, 89

mental health, 45, 52

mentality, 1, 3, 4, 17, 20, 24, 27, 29, 35, 36, 37, 38, 40, 41, 44, 84, 93, 94, 99, 100, 105, 175, 178, 179, 180, 182, 223, 242, 248

meta information, 71

meta-analysis, 204, 280, 283

metabolic engine, 207

metabolic rate, 165
metabolic syndrome, 231
metabolically healthy, 56
metabolise, 157
metabolism, 4, 6, 44, 47, 81, 87, 91, 93, 95, 97, 122, 161, 164, 188, 189, 203, 217, 221, 222, 233, 241, 242, 281, 284
metaphor, 9
metaphorical ogre, 100
middle-aged deaths, 169, 171
middle-eastern, 189
milk stout, 222
imicking a Gut Hormone, 237
mindfulness Interventions, 242
mindless Eating, 134, 284
mindset, 1, 2, 3, 4, 6, 32, 81, 175, 247
minerals, 78, 168, 170, 234
minimal effort., 40
mini-meals, 197
mirror, 25, 32, 41, 43, 46, 47, 48, 50, 51, 52, 54, 60, 61, 111, 140, 141, 142, 176, 177, 247
mirror test, 52
misguided beliefs, 184
misinformation, 27
misinterpretation, 37
moderately overweight, 66
money, 3, 7, 11, 25, 26, 38, 40, 45, 50, 79, 83, 86, 184, 186, 192

moon face, 98
morbidity, 7, 25, 68, 228
morbidly obese, 252
mortality, 7, 25, 68, 74, 77, 198, 204, 228, 229, 252, 280, 281, 282, 283
motivate, 61, 219
motivation, 21, 32, 45, 49, 61, 70, 176, 177, 187, 242
motivational interviewing, 242
motto, 183
muscle and joint pains, 207
muscle wasting, 98
muscles grow, 220
musical, 14, 15
myth, 9, 222

N

naïve, 179
Napoleon, 23
negative, 11, 46, 113, 134
negotiable, 157
neurochemicals, 6, 95
no love sincerer, 118
'no pain, no gain', 80
non-compliance, 192
nor-epinephrine, 240
nostalgia, 19
Nuremberg defence, 237

nutrient content, 153, 218

nutrients, 6, 167, 168, 169, 170, 172, 173, 243, 266

nutrition, 7, 157, 226, 234

Nutrition Examination Surveys, 78, 283

nutritional guidance, 87

nutritionist, 27

nuts and seeds, 172, 234

O

oatmeal, 161

obese, 26, 45, 55, 64, 70, 75, 76, 83, 84, 87, 94, 110, 160, 204, 229, 237, 243, 249, 252, 283, 284

obesity, 3, 7, 11, 22, 25, 27, 46, 55, 56, 57, 58, 69, 70, 72, 73, 74, 75, 76, 78, 82, 83, 84, 89, 90, 93, 97, 98, 107, 119, 128, 134, 145, 186, 188, 189, 198, 200, 202, 226, 228, 233, 238, 239, 246, 249, 255, 280, 281, 284

obsessional, 122

obsessive trait, 198

oestrogen, 98

offal, 170, 172, 174

optical illusion, 62

opulence, 11

organic disease., 52

Orlistat, 239

oscillate, 57

overactive thyroid, 98

overconsumption, 128

overeating, 36, 92, 189

over-indulgence, 57, 176

overweight, 2, 4, 11, 13, 15, 18, 21, 25, 26, 30, 32, 44, 45, 46, 47, 51, 54, 55, 56, 57, 59, 60, 61, 63, 65, 68, 69, 70, 72, 73, 75, 76, 81, 87, 88, 91, 94, 96, 97, 100, 106, 107, 110, 120, 121, 122, 128, 133, 137, 158, 175, 178, 180, 185, 189, 195, 196, 197, 199, 203, 204, 205, 206, 217, 221, 222, 227, 229, 242, 243, 245, 246, 247, 249, 252, 281

overweight people, 2, 4, 5, 25, 30, 32, 45, 46, 51, 70, 72, 81, 96, 121, 122, 137, 197, 199, 204, 205, 222, 229, 249

oxidative, 165

oxygen, 62, 64, 169, 204, 207

Ozempic®, 237, 238

P

palpitation, 98, 235, 255

parable, 9

parents, 123

pastoral care, 88

paternalistic, 72, 88

patients, 4, 44, 59, 64, 69, 71, 72, 73, 80, 84, 88, 97, 110, 120, 160, 169, 171, 184, 191, 214, 235, 237, 238, 239, 243, 284

patisseries, 162

peptide YY, 96

personal, 3, 18, 27, 40, 44, 51, 55, 71, 88, 99, 100, 176, 213, 242

personal responsibility, 27, 176

personal trainer, 27

personality, 44

pharmaceutical companies, 22
phendimetrazine, 236
phentermine, 235
philanthropy, 40
philosophy, 14
phobias, 123
physical activity, 3, 23, 25, 191, 204, 283
physical endeavour, 99
physical fitness, 4, 13, 30, 43, 166, 184
physical prowess, 83
physically ill, 156
physicians, 230
physiological problem, 62
pinch an inch, 56, 66, 67
pituitary gland, 98
placebo, 238, 239, 240, 281
plaques (arterial), 169
pleasure and leisure, 39
pleasure principle, 122
podgy, 45
policy, 17
political, 7, 135, 228
political perspective, 7
politicians, 228
politicians, 72
polyphenols, 224

poorly motivated, 107

populations, 110

portion sizes, 3, 86, 121, 134, 135, 139, 146, 147, 152, 162, 246

possessions, 12, 18

potassium, 226, 234

potential partners, 111

poverty, 7, 25, 26, 40, 57, 74, 78, 228

prebiotics, 232, 280

predicting disease, 89

prehistoric humans, 82, 163

prehistoric life, 187, 202

presumption, 60

prevalence, 7, 119, 186, 202

preventing heart attacks, 172

pride, 10, 214

probiotics, 232, 233, 280

professional help, 146, 179

profit, 40, 184

progesterone, 98

progress, 23, 50, 107, 113, 132, 135, 136, 140, 142, 206, 207, 221

progression, 254

protective substance, 170

protein, 7, 19, 115, 141, 156, 157, 158, 160, 162, 166, 183, 218, 220

psychiatric condition, 198

psychiatry, 230

psychobabble, 49

psychological, 2, 4, 48, 69, 113, 192, 230, 237, 241, 242, 248

psychological Interventions, 241

psychology, 6, 14, 93, 187

puberty, 98

pyramid, 102, 103, 104, 105, 112, 113, 116, 133, 155

Pyramid of the Sun, 105

Q

quality of their life., 89

Queen, 111, 194

Queen Mother, 111

questionnaire, 28, 33, 37, 182, 209

questions, 22, 23, 44, 80, 85, 105, 109, 137, 164, 205, 210

R

rationing, 135

recipes, 163

reduced calorie, 151

reflexive, 29

regimes, 1, 52, 86, 213

re-iteration, 177

relationship, 45, 55, 74, 109, 184, 227

relationships, 3, 78, 198

relaxed, 10

relevance to your health, 54

reliable guide, 46, 77

religious observances, 123

repairing the cells, 157
repeated failure, 81, 106
repetition, 140, 207
reserves, 25, 26, 58, 221
resign, 32, 36
restaurants, 10, 14, 30, 39, 83, 119, 134, 183, 190, 192
retreat, 23
reverence, 39, 143, 183
reverence for food, 143
rich, 7, 11, 17, 22, 23, 25, 26, 40, 59, 83, 138, 145, 157, 162, 163, 185, 199, 224, 227, 228, 229, 246
ripple, 56, 57, 58, 66
risk factors, 110, 282
risks, 45, 52, 70, 255
robots using AI, 202
role, 1, 3, 21, 81, 125
roughage, 128
rowing, 203
rules, 13, 30, 113, 186, 190, 193, 194
Rumsfeld, 13
Russian military, 241
S
salt, 172, 173, 219, 225, 226, 227, 229, 234, 235, 284
salt and water, 227
sanctimonious, 83
satiation, 47, 95, 128, 151, 181, 233, 235, 243

satiety, 96

satiety hormone, 96

saturated fat, 77, 160, 167, 168, 171, 172

Saxenda®, 239

schools, 17

science, 1, 13, 72, 81, 111, 120, 216

sea swimmers, 63

Second World War, 59, 123, 134, 158, 194, 198, 236

secret formula, 43

secret way, 50

security, 25, 40

sedentary people, 135, 217

seduced, 85

selenium, 170, 234

self-conscious, 31

self-control, 128

self-deception, 106

self-discipline, 32, 81, 99, 151, 213, 215, 223

self-esteem, 5, 12, 25, 36, 47, 48, 51, 99, 100, 110, 112, 143, 184, 185, 214, 228

self-fulfilment, 9

self-gratification, 10

self-indoctrination, 177

self-indulgence, 2, 13, 18, 119, 188

selfishness, 112, 185

semaglutide, 237, 238, 239, 247

sense of failure, 145

sensitive, 31, 172

serotonin, 92

servants, 11, 14

setmelanotide, 239

seven-day, 142, 151

shape and weight, 43, 48, 60, 92

shellfish, 17, 170, 171, 172

shopping, 10

Shopping, 40, 144

shops, 10, 23, 119, 163, 231

short-cuts, 108

shorter life, 52, 249

shortness of breath, 238, 255

shy person, 32, 194

sibutramine, 240

six-pack, 222

skills, 13, 15, 185

sleep, 191, 221, 238, 244

sleeve gastrectomy, 243, 244

slightly overweight, 67

slim, 2, 11, 26, 30, 32, 64, 69, 83, 120, 122, 180, 182, 185, 186, 187, 188, 195, 203, 227, 237, 248, 249

slimming pills, 197

sluggish thyroid, 97

small intestine, 95, 96, 243, 244

smelly breath, 165
smoking, 7, 25, 37, 49, 86, 87, 89, 105, 128, 214, 228, 255
social circles, 193
social problems, 228
social support interventions, 242
socially advantaged, 7
societal role, 70
society, 11, 13, 45, 73, 186
socio-economic status, 228
sodium chloride, 225
Spartans, 10, 83, 84
spending, 40
spiralina, 231
sports, 21, 63
stamina, 188, 245
starch, 138, 156, 158, 225
starvation, 95, 124, 192, 221
statistical evidence, 73
statistical paradox, 55
statistics, 46, 55, 70, 71, 72, 74, 249, 252
step pyramid, 108
Stone Age, 163
stored fat, 58, 82, 158
strain, 206, 207
strategy, 40, 88, 96, 118, 196, 199
stress, 7, 10, 51, 52, 81, 106, 126, 143, 165

stressed and depressed, 105

strict rules, 190

strokes, 6, 25, 46, 72, 110, 167, 172, 238

struggle, 80, 86, 135, 182

sugar, 95, 123, 138, 144, 158, 163, 173, 218, 223, 224, 225, 231

Sumo wrestlers, 63

supermodel, 111

supplementation, 170, 171

suppress, 38, 146, 243

suppressant, 22, 118, 121, 237, 282

surgeons, 230

surgical intervention, 242

survival, 37, 45, 74, 82, 94, 105, 157, 187, 221, 228

survival of the fittest, 45, 82, 187

survive healthily, 37

sweating, 241

sweet things, 159, 164

swimming, 10, 57, 63, 65, 92, 203, 204

swollen feet, 255

symptoms, 96, 97, 169, 254

synbiotics, 232, 233, 280

T

T4 and T3, 97

tables of data, 60

take it easy, 24, 27, 40, 166, 178

talk to ourselves, 29

taurine, 170, 171

taxing, 11, 64, 246

technical information, 59

Teddy Boys, 214

Thinking Fast and Slow, 29, 283

thyroid-stimulating hormone, 98

Tom Ketteridge, 163

top killer in the western world, 73

tough-minded, 46

trans-fat, 168, 169, 171, 173

trivial pursuits, 14

trying to lose weight, 6, 34, 85, 86, 96, 101, 120, 156, 197

tummy, 56, 57, 141, 177, 222

twins, 92, 93

twins, 284

Type A Behaviour, 196

U

underweight, 11, 20, 47, 56, 57, 59, 63, 68, 69, 77, 87, 189, 198, 223, 252, 255

unfit, 38, 82, 180, 181, 207, 220, 221, 222

unhappy, 26, 30, 48, 51, 52, 54, 105, 119, 126, 248

unhealthy pursuit, 48

university degrees, 13, 17

unknown, unknown, 13

unnatural, 22, 48

urine for ketones, 164

V

vanity, 24, 110, 111

variation, 94

vegetables, 12, 17, 77, 83, 145, 170, 234

vibrate, 57, 58

vicious cycle, 22

Victoza®, 239

vigilance, 32

visible ribs, 67

vitamin C, 224

vitamins, 78, 157, 168, 170

voice in your head, 99, 127

volume, 16, 62, 63, 135, 148, 151

W

waist size, 56, 59

walking, 10, 50, 56, 57, 63, 65, 66, 67, 99, 193, 202, 203, 204, 205, 206, 245, 246

waste, 10, 51, 118, 183

water, 57, 92, 118, 124, 127, 128, 140, 144, 161, 168, 174, 199, 220, 224, 226, 227, 234, 235

weakness, 124, 238

wealth, 18, 25, 40, 57, 184, 228

Wegovy®, 237, 238, 282

weighing scales, 149, 219

weight gain, 2, 10, 19, 25, 32, 45, 57, 62, 63, 64, 87, 96, 97, 98, 156, 158, 191, 198, 220, 221, 222, 223, 225, 226, 233, 235 weight loss, 1, 2, 3, 6, 22, 24, 25, 30, 31, 43, 44, 48, 49, 52, 55, 61, 73, 79, 80, 81, 85, 86, 89, 95, 96, 98, 105, 106, 107, 108, 110, 112, 113, 114, 120, 121, 151, 161, 164, 165, 173, 181, 185, 203, 204, 209, 213, 216, 217, 218, 219, 220, 221, 222, 223, 224, 227, 229, 230,231, 232, 233, 234, 237, 239, 240, 241, 242, 243, 245, 246, 247

weight problems, 50, 88

weight reduction, 60, 120, 281

weight training, 203, 204, 206, 207, 208, 221, 222

weight-gain, 22

weight-gaining potential, 145, 200

weights under tension, 204

well-rounded, 10, 45

western countries, 7

western cultures, 57

western societies, 24, 83

western tradition, 189

wheat, 163

white wine, 153, 161

wisdom, 113

Wizard of Oz, 49

wizards, 50

wobble, 56, 57, 58, 61, 64, 67

working-class, 84

Y

yang, 84

ying, 84

younger self, 5, 46

Z

Zagluton, 18, 19

zinc, 170, 234

Zobesia, 9, 11, 13, 14, 17, 18, 19, 21, 23, 41

Zobesian, 10, 11, 13, 14, 15, 16, 17, 18, 19, 20, 32, 166

www.ingramcontent.com/pod-product-compliance
Lightning Source LLC
Chambersburg PA
CBHW070759040426
42333CB00060B/942